A LITTLE BOOK OF PACKAGING CONCEPTS

Authored by

Nishant Sharma

Contents

Preface

Packaging is far more than a protective layer around a product—it is a silent communicator, a brand ambassador, and often the first point of interaction between a product and its user. In today's fast-paced and highly competitive marketplace, packaging plays a critical role not only in preserving and transporting goods but also in influencing purchasing decisions, ensuring safety, and delivering convenience.

A Little Book of Packaging Concepts is designed to serve as a concise yet meaningful guide for students, professionals, and anyone curious about the world of packaging. This book brings together fundamental ideas, practical insights, and essential principles that form the backbone of modern packaging practices. Rather than overwhelming the reader with complexity, it aims to simplify key concepts and present them in an easy-to-understand manner.

Whether you are new to the field or have years of experience, revisiting the basics often reveals new perspectives. From material selection and design considerations to sustainability and regulatory aspects, each concept in this book contributes to building a strong foundation in packaging knowledge.

This book is not intended to be exhaustive, but rather a starting point—a small companion that encourages deeper exploration and continuous learning. Packaging is an evolving discipline, shaped by innovation, consumer needs, and environmental responsibility. It is my hope that this book inspires readers to think critically, design responsibly, and contribute meaningfully to the future of packaging.

About the Author

Nishant Sharma is a seasoned Packaging Technologist with over 20 years of comprehensive experience in packaging design, development, quality assurance, costing, procurement, logistics, technical sales, and business development. His career spans across diverse industries including food, consumer goods, automotive, and specialty materials, giving him a well-rounded perspective on packaging as both a science and a strategic business tool.

He holds a Post Graduate Diploma in Packaging from the Indian Institute of Packaging (IIP)-Delhi and has a deep passion for simplifying complex packaging concepts into practical knowledge that can be easily understood and applied.

Through A Little Book of Packaging Concepts, Nishant aims to share his experience and insights with students, young professionals, and industry peers, encouraging them to build strong fundamentals and approach packaging with curiosity, responsibility, and innovation.

First Edition: 2026

Published by:
Nishant Sharma

Cover design and layout by the author.

For permissions, inquiries, or feedback, please contact: nsharma.iip@gmail.com

Printed in India

Chapter Structure

No.	Title	Strategic Purpose
1	Introduction to Packaging Concepts	Orient readers, establish relevance, and set learning objectives.
2	Defining Packaging	Clarify scope, functions, terminology, and industry standards.
3	Materials and Their Properties	Present core material types and their performance attributes.
4	Types of Packaging	Differentiate primary, secondary, and tertiary packaging levels.
5	Packaging Design Principles	Teach structural and graphic design fundamentals, build skills.
6	Manufacturing Processes	Explain key converting, forming, filling, and sealing methods.
7	Testing and Quality Assurance	Introduce testing protocols, compliance, and quality systems.
8	Sustainability and Regulations	Explore environmental impact, circular economy, and regulations.
9	Careers in the Packaging Industry	Guide professional paths, required skills, and career planning.
10	Future Trends and Next Steps	Highlight innovations, smart packaging, and resources ahead.
11	Limitations and Challenges in Packaging Design	Real-world packaging constraints, balanced solutions.
12	Practical Insights and Real-World Case Studies	Highlighting practical decision-making and setting learning objectives for applying concepts in industry scenarios.

Chapter 1: Introduction to Packaging Concepts

The significance of packaging extends far beyond the humble container that holds a product. From the moment a design sketch is translated into material form, packaging becomes a dynamic nexus of protection, communication, and logistics. For students and emerging professionals, understanding the multifaceted role of packaging lays the groundwork for effective design decisions and strategic problem solving. In this chapter, we will examine how packaging functions to safeguard goods, enhance brand value, and streamline supply chains. We will trace its evolution through history and conclude by exploring its contemporary impact on business, society, and the environment.

The Role of Packaging

At its core, packaging serves four primary functions: protection, containment, communication, and convenience. First and foremost, packaging shields products from physical damage, microbial contamination, and environmental stresses. Whether safeguarding a fragile glass vial or preserving the freshness of a perishable food item, the structural integrity of packaging materials must be carefully matched to the product's vulnerabilities. Material selection, thickness, and design geometry all play critical roles in mitigating shock, vibration, moisture, and temperature fluctuations during transportation and storage.

Beyond protection, packaging provides effective containment, holding products in a defined shape and volume for handling and display. Containment simplifies handling for manufacturers, retailers, and consumers alike, enabling efficient filling, stacking, and shelving. Furthermore, packaging communicates vital information: ingredients, usage instructions, safety warnings, and branding messages. The choice of graphic elements, typography, and color palette must align with regulatory requirements and brand identity while remaining legible and engaging for the target audience.

Finally, convenience features—easy-open tabs, resealable closures, and portion control devices—enhance user experience and encourage repeat purchase. By anticipating the needs and behaviors of end users, packaging designers can create intuitive interactions that reduce frustration and reinforce brand loyalty. Effective packaging design thus unites engineering, marketing, and human factors to deliver solutions that satisfy product requirements and delight consumers.

Historical Evolution

Packaging has accompanied human civilization from its earliest days. Primitive containers—woven baskets, animal skins, hollowed gourds—provided simple means of transporting grains, liquids, and other essentials. These organic vessels reflected the materials at hand and the ingenuity of their makers. With the advent of metallurgy, metal vessels and tins emerged, offering greater durability and reusability for commodities such as oils, preserves, and paints.

The Industrial Revolution ushered in mass production of packaging materials. Tinplate cans enabled the wide-scale preservation of perishable foods, fuelling urban growth and global trade. Paperboard cartons appeared in the late nineteenth century, initially as protective inner liners and later as primary enclosures for consumer goods. Corrugated fiberboard, patented in the early twentieth century, transformed distribution logistics with its exceptional strength-to-weight ratio, permitting lightweight pallets and standardized shipping cases.

In the post-war era, the rise of plastics revolutionized the industry once more. Polymers such as polyethylene and polypropylene allowed for flexible films, blow-molded bottles, and thermoformed trays that were lightweight, moisture-resistant, and highly versatile. Simultaneously, advances in printing technology enabled high-resolution graphics, pushing packaging to the forefront of brand communication. Today, packaging has evolved into an integrated system, balancing traditional materials with innovative substrates and smart technologies to meet the complex demands of global markets.

Importance in Modern Industry

In the twenty-first century, packaging stands as a vital economic driver. It represents a significant share of manufacturing output and contributes directly to the competitiveness of consumer-facing goods. Efficient packaging design reduces material costs, optimizes transportation efficiency, and minimizes waste, delivering measurable savings at every stage of the supply chain. Companies that master packaging innovation often gain a strategic edge, whether through premium unboxing experiences or by reducing their carbon footprint.

Environmental considerations now occupy a central role in packaging decision making. Life cycle thinking prompts designers to assess raw material sourcing, production energy, recycling potential, and end-of-life disposal. Lightweighting initiatives reduce resource consumption and greenhouse gas emissions, while advances in compostable biopolymers and closed-loop recycling systems offer pathways toward circular economy models. Packaging professionals must stay informed about evolving regulations, consumer expectations, and technology developments that shape sustainable strategies.

From the consumer perspective, packaging influences purchasing behavior and brand perception. A well-designed package not only protects the product but also conveys authenticity, quality, and value. Clear labeling builds trust, while thoughtful ergonomics and convenience features enhance satisfaction and usability. In an era of e-commerce, packaging must also address unique challenges—protecting goods during transit, managing returns, and creating memorable unboxing experiences that foster social media engagement.

Conclusion

By exploring the protective, communicative, and logistical roles of packaging, its historical transformation, and its contemporary economic and environmental impacts, we have established a conceptual framework for understanding this vital discipline. As you move forward into deeper discussions of materials, design principles, and industry practices, keep in mind the foundational interactions between product requirements, user needs, and system efficiencies. This holistic perspective will guide your decisions and innovations as you progress through the world of packaging.

CHAPTER 1

Introduction to Packaging Concepts

Packaging is more than a container—it is a critical link between products and people.

PROTECTS | COMMUNICATES | SUSTAINS | DELIVERS VALUE

Purpose of this chapter:
Orient readers, establish relevance, and set learning objectives.

1. ORIENT READERS

Introduce the fundamental ideas, terminology, and scope of packaging concepts.

- What packaging is—and what it is not
- Key functions and components
- Types and levels of packaging
- A systems perspective

2. ESTABLISH RELEVANCE

Highlight why packaging matters across industries and everyday life.

- Protects products and ensures safety
- Influences consumer choices
- Supports logistics and convenience
- Drives sustainability and reduces impact
- Creates brand value and differentiation

3. SET LEARNING OBJECTIVES

Prepare readers for what they will learn in this book.

- Understand core packaging concepts
- Evaluate materials and structures
- Analyze performance and trade-offs
- Apply concepts to real-world solutions
- Think critically and innovatively

By the end of this chapter, you will have a clear foundation for exploring how packaging meets needs, solves challenges, and shapes a better future.

Chapter 2: Defining Packaging

Definition and Scope

Packaging, in its broadest sense, encompasses the materials and processes that encase, protect, and present products throughout their lifecycle. At a fundamental level, it is the interface between a product and the world—providing a barrier against physical forces, environmental conditions, and contamination while enabling handling, storage, and display. Yet this simple conception belies the complexity inherent in defining packaging for an industry that spans raw materials, converting operations, distribution networks, retail environments, and consumer use.

To establish a working definition, consider packaging as a multidisciplinary system comprised of materials (paper, board, plastics, metals, glass, and emerging substrates), structures (boxes, bottles, pouches, trays, and their myriad combinations), and ancillary elements (closures, labels, protective inserts). This system fulfills functions that extend well beyond containment: it shapes consumer perception, facilitates logistics, and influences a product's environmental footprint. The scope of packaging therefore ranges from the primary enclosure that directly contacts a product—such as a chocolate bar's foil wrapper—to secondary groupings designed for merchandising—like a paperboard tray holding multiple bars—and tertiary solutions for bulk shipment—pallet configurations wrapped in stretch film.

Industry classification further clarifies packaging's boundaries. Primary packaging comprises the immediate interface with the product: it must meet stringent requirements for food safety, child resistance, or chemical compatibility. Secondary packaging bundles one or more primary units, often optimizing shelf display, brand communication, or transportation efficiency. Tertiary packaging addresses large-scale distribution, focusing on load stability, palletization standards, and handling equipment compatibility. Each level operates within its own design constraints and performance criteria, yet must integrate seamlessly into the broader packaging ecosystem.

By defining packaging as a layered system of enclosures and supports, we recognize its dual nature: a technical discipline concerned with material science, mechanics, and regulatory compliance, and a creative endeavor that draws on graphic design, marketing strategy, and human factors. For beginners entering the field, this definition underscores the importance of interdisciplinary collaboration—engineers, designers, supply-chain specialists, and sustainability experts must align their perspectives to craft packaging solutions that meet functional, economic, and environmental objectives.

Core Functions

Every packaging design emerges from an understanding of four core functions: protection, containment, communication, and convenience. These pillars guide material selection, structural design, and graphic treatment, ensuring that a package fulfills its purpose from factory line to end user.

Protection lies at the heart of packaging engineering. A robust package mitigates hazards encountered during transportation and storage—shock from drops, vibration during transit, temperature extremes, and exposure to moisture or gases. Engineers employ mechanical tests such as drop trials and vibration profiling to quantify protective performance, then iterate structural designs and material choices accordingly. Selecting the appropriate board grade, plastic film thickness, or engineered foam insert can mean the difference between intact product delivery and costly damage claims.

Containment refers to a package's ability to hold a defined quantity of product securely and maintain its integrity. Liquid goods demand leak-proof seals; granular solids require barrier properties against moisture ingress; powders may call for inner liners to prevent sifting. Beyond physical retention, containment organizes products for efficient handling—standardized container sizes facilitate automated filling lines, pallet configurations, and retail shelving. By imposing consistent form factors, the containment function simplifies processes on both manufacturing and distribution fronts.

Communication transforms packaging into a silent salesperson. Through graphics, typography, color palettes, and structural cues, packaging conveys brand identity, regulatory information, usage instructions, and sustainability credentials. A well-conceived label hierarchy ensures that critical data—ingredients, hazard warnings, nutritional facts—stands out legibly, while visual elements reinforce brand messaging. Successful packaging design leverages this communicative power to captivate consumers, differentiate products on crowded shelves, and build lasting emotional connections.

Convenience addresses the end-user experience, anticipating how consumers open, pour, dispense, reseal, and store products.

Ergonomic handles, tear-notches, reclosable caps, and portioned servings all contribute to user satisfaction. A poorly designed closure that ruptures or sticks will erode brand loyalty, whereas intuitive opening features and resealable mechanisms can elevate perceived value. By studying consumer behavior and testing prototypes in real-world scenarios, designers refine convenience elements that foster repeat purchase and positive reviews.

Economic and Consumer Perspectives

Packaging decisions seldom rest solely on technical performance; they are shaped equally by cost considerations and consumer expectations. On the economic side, material and processing costs, production speed, waste rates, and logistics expenses combine to determine the total cost of packaging. Manufacturers strive for economies of scale—leveraging high-speed filling lines and large-volume material orders—while balancing the trade-off between premium materials and budget constraints. Lightweighting strategies, which reduce material usage without sacrificing strength, exemplify cost-saving approaches that also support sustainability goals.

From the consumer's standpoint, packaging is both functional necessity and brand ambassador. Shoppers often judge product quality based on the look and feel of its packaging. Premium finishes, tactile coatings, and innovative structures can command higher price points, whereas inexpensive or flimsy materials may undermine consumer confidence. Additionally, today's buyers are increasingly attuned to environmental claims: recyclable content, minimal waste design, and evidence of third-party certifications influence purchase decisions. Packaging professionals must therefore align material choices and messaging with evolving market trends, ensuring that economic imperatives and consumer desires reinforce rather than contradict one another.

Terminology and Standards

A working fluency in packaging terminology and industry standards is essential for clear communication and regulatory compliance. Terms such as "gross weight," "tare," "net weight," and "fill line" describe the mass characteristics of packaged products, while "barrier properties," "seal strength," and "burst resistance" quantify material performance. Understanding structural descriptors—like "flute profile" in corrugated board or "molecular orientation" in plastic films—enables precise specification and testing.

Standards established by organizations such as ISO (International Organization for Standardization) and ASTM (American Society for Testing and Materials) provide common frameworks for packaging quality and safety. ISO 11607, for example, outlines requirements for packaging materials used in medical device sterilization, whereas ASTM D4169 specifies performance testing for distribution packaging. Compliance with these standards ensures that packaging meets globally recognized benchmarks for durability, hygiene, and environmental resistance.

Beyond international norms, regional regulations—such as the European Union's Packaging and Packaging Waste Directive—dictate recovery targets and labeling requirements for recyclable content. Food-contact materials must comply with local food safety authorities' mandates on migration limits and approved substances. In the pharmaceutical sector, child-resistant closures and tamper-evident features are subject to stringent guidelines. By mastering terminology and adhering to applicable standards, packaging professionals can navigate the regulatory landscape with confidence, avoiding costly redesigns and ensuring that products reach markets without delay.

In this chapter, we have defined packaging as a system of layered enclosures that perform protection, containment, communication, and convenience functions, all within economic and regulatory frameworks. We have outlined the economic and consumer forces that shape material and design choices, and we have introduced key terminology and standards that govern industry practice. Equipped with this conceptual foundation, you are prepared to explore the materials, structures, and processes that comprise the rich discipline of packaging design and innovation.

CHAPTER 2

Defining Packaging

What It Is, What It Does, and How We Talk About It

Packaging is the coordinated system of preparing goods for storage, transport, sale, and use. It encompasses more than containers—it includes materials, structures, information, and processes that work together to deliver value.

THE SCOPE OF PACKAGING

From the point of manufacture → Through the supply chain → At the point of sale → To the point of use → And beyond (end of life)

Purpose of this chapter:
Clarify scope, functions, terminology, and industry standards.

KEY FUNCTIONS OF PACKAGING

- Protects
- Contains
- Communicates
- Facilitates handling and transport
- Supports marketing
- Enhances convenience
- Promotes sustainability

COMMON TERMINOLOGY

Package: The total system for containing, protecting, handling, delivering, and presenting products.

Packaging Material: Any substance used in or for a package.

Container: A receptacle in direct contact with the product.

Pack: The complete packaged product ready for distridution or sale.

Unit Load: A number of packs or packages assembled for handling and transport.

INDUSTRY STANDARDS AND REFERENCES

- **ISO 9001** – Quality management
- **ISO 14001** – Environmental management
- **ISO 18601** – Packaging and the environment
- **ASTM standards** – Materials and test methods
- **ISTA standards** – Package testing for safe transport
- Regional regulations and guidelines (e.g., FDA, EU Packaging Directive)

TYPES OF PACKAGING (LEVELS)

Primary Packaging
In direct contact with the product

Secondary Packaging
Groups primary packages together

Tertiary Packaging
Protects and facilitates transport of grouped packages

By the end of this chapter, you will be able to:
Define packaging, explain its functions, use key terms correctly, and recognize relevant standards that guide the industry.

Chapter 3: Materials and Their Properties

In packaging design, the choice of material forms the foundation upon which protection, functionality, and sustainability are built. Each substrate—whether paperboard, plastic, metal, or glass—brings a distinctive balance of mechanical strength, barrier performance, cost, and environmental impact. In this chapter, we explore four principal material families: paper and board, plastics, metal and glass, and emerging biopolymers. By understanding their intrinsic properties and application nuances, you will gain the insight needed to select or innovate materials that align with product requirements and market demands.

Paper and Board

Paper and board materials dominate primary and secondary packaging in sectors ranging from confectionery to e-commerce. Their appeal lies in light weight, printability, recyclability, and the well-established recycling infrastructure. Yet within this family exist numerous grades and structures:

Single-face corrugated board combines a flat liner and a fluted medium to yield cushioning strength with minimal weight. When two liners sandwich two flutes, double-wall board can support heavy loads on pallets. The flute profile—A, B, C, or E—determines stacking strength and crush resistance.

Solid bleached sulfate (SBS) paperboard, made from virgin pulp, provides a pristine, high-brightness surface ideal for premium graphic finishes. Folding carton board, often used for cereal boxes or cosmetic cartons, balances stiffness with foldability, enabling intricate die-cuts and window features.

Barrier coatings, such as polyethylene or food-safe lacquers, extend board utility by imparting moisture resistance. Designers must consider coat weight and bonding quality; too thin, and liquids seep through; too heavy, and recyclability suffers.

When specifying paper and board, evaluate burst strength (resistance to pressure), edge crush test (ECT, for stacking), and grammage (mass per square meter). Pair these metrics with moisture-vapor transmission rate (MVTR) if contents are sensitive to humidity. By tuning thickness, flute structure, and coatings, paper and board can cost-effectively satisfy protection and communication needs.

Plastics

Plastic materials offer unmatched versatility in film, sheet, and rigid form. Their barrier properties, transparency, and thermoforming potential make them staples in food, pharmaceutical, and consumer electronics packaging. Key polymer families include:

Polyethylene (PE), available as low-density (LDPE) or high-density (HDPE), lends itself to flexible films, squeeze bottles, and liners. Its moisture barrier and sealability are excellent, though gas permeability can be high.

Polypropylene (PP) combines stiffness with heat resistance, making it suitable for hot-fill pouches, injection-molded caps, and rigid containers. Oriented PP film (OPP) enhances tensile strength and clarity for snack packaging.

Polyethylene terephthalate (PET) dominates clear beverage bottles, offering superior clarity, strength, and oxygen barrier, especially when combined with coating or barrier layers.

Multilayer structures—coextruded or laminated—pair polymers to achieve tailored barrier and mechanical profiles. A three-layer film, for instance, might sandwich an ethylene vinyl alcohol (EVOH) barrier core between PE layers to protect oxygen-sensitive foods.

While plastics deliver performance, they pose recycling and sustainability challenges. Resin identification codes guide material separation, but mixed formats and barrier layers complicate recycling streams. Emerging design for recyclability focuses on mono-material solutions or easily delaminated layers. When selecting plastic, weigh recyclability initiatives, consumer expectations, and evolving regulations on single-use plastics.

Metal and Glass

Metal and glass bring high barrier performance and premium consumer perception to packaging, albeit at higher material cost and weight. Glass, composed of silica, soda ash, and lime, forms impervious containers that resist gas and moisture migration indefinitely. Its inertness is ideal for food, beverage, and pharmaceutical applications where flavor integrity and chemical stability are paramount. The familiar "pop" of a glass jar conveys quality, but glass's brittle nature demands careful wall-thickness design and protective secondary packaging to prevent breakage during transit.

Aluminum and tinplate steel serve as metals of choice for cans, closures, and foil. Aluminum's light weight and formability allow for beverage cans and aerosol containers; its protective oxide layer offers mild corrosion resistance. Tinplate steel, coated with tin or chromium, finds use in food cans and closures. Both metals excel in barrier function and infinite recyclability—aluminum scrap can return to the production line repeatedly. Yet designers must account for metal creep during can stacking and potential corrosion if coatings fail. Proper material thickness, coating integrity, and end-of-line quality inspection ensure reliable performance.

Emerging Materials

As sustainability moves from trend to imperative, novel substrates are reshaping the packaging landscape. Biopolymers derived from plant starches, polylactic acid (PLA), or cellulose offer compostable alternatives to conventional plastics. PLA, produced from fermented corn sugar, yields transparent films and rigid containers that biodegrade under industrial composting conditions. Cellulosic films—made from wood pulp—provide excellent grease resistance and compostability but may require barrier coatings that compromise end-of-life options.

Paper-plastic composites using water-soluble adhesives facilitate separation of layers for recycling. Mushroom and mycelium-based foams mimic polystyrene's cushioning with fully compostable profiles. Algae-derived polymers are under development for thin films. Each emerging material demands rigorous supply-chain vetting, performance testing, and clear labeling to manage consumer expectations about disposal.

When evaluating emerging materials, focus on certifications (e.g., EN 13432 for compostability), industrial-compost vs. home-compost distinctions, and local waste-management capabilities. Piloting new substrates on small production runs can uncover processing challenges—extrusion temperatures, seal-strength calibration, or print adhesion—that differ from conventional materials. By embracing innovation while rigorously testing performance, packaging professionals can transition toward circular-economy models without compromising product integrity.

Bringing It All Together

Selecting the right material involves balancing protection, barrier properties, sustainability, cost, and consumer perception. The process begins with a clear statement of product requirements: Is the content sensitive to oxygen, moisture, or light? Must the package withstand high-speed filling and sealing operations? What disposal or recycling pathways must the design support? Armed with precise performance targets, you can compare material candidates using standardized tests—MVTR, oxygen transmission rate (OTR), drop and compression trials—and lifecycle assessments.

Case in point: a refrigerated ready-meal tray might combine a PP thermoformed base, a laminated barrier film lid, and a paperboard sleeve for branding. PP provides rigidity and heat resistance; the barrier film protects against oxygen; the sleeve offers graphic real estate and a familiar paper feel. Each component is chosen to meet functional needs while allowing separate recycling streams. This hybrid solution exemplifies how material properties coalesce into a cohesive packaging system.

As you progress in packaging design, maintain a dialogue with material suppliers, converters, and waste-management experts. Testing prototypes under real-world conditions—transport simulation, consumer use trials, end-of-life trials—uncovers subtle interactions that laboratory data alone may miss. By combining material science rigor with practical experimentation, you will develop packaging that not only safeguards and communicates but also advances sustainability and resonates with today's eco-conscious consumers.

CHAPTER 3

Materials and Their Properties

Present core material types and their performance attributes.

Materials are the foundation of engineered products and systems.
Each material type has unique properties that determine how it performs in different applications.

METALS

Strong, tough, and durable materials with excellent structural performance.

KEY ATTRIBUTES

- High strength
- Toughness
- Good thermal and electrical conductivity
- Generally ductile and malleable
- Susceptible to corrosion (varies)

POLYMERS

Lightweight and versatile materials with a wide range of applications.

KEY ATTRIBUTES

- Low density
- Good chemical resistance
- Low thermal conductivity
- Moderate strength and stiffness
- Can be molded easily; recyclable (varies)

CERAMICS & GLASSES

Hard, brittle materials with excellent wear and thermal resistance.

KEY ATTRIBUTES

- Very hard
- High temperature stability
- Excellent wear and corrosion resistance
- Electrical insulators
- Brittle; low tensile strength

COMPOSITES

Engineered combinations of materials designed for high performance.

KEY ATTRIBUTES

- High strength-to-weight ratio
- High stiffness
- Good corrosion resistance
- Tailorable properties for specific needs
- Often anisotropic (direction-dependent)

NATURAL MATERIALS

Derived from natural sources; renewable and often sustainable.

KEY ATTRIBUTES

- Renewable
- Biodegradable
- Low thermal conductivity
- Variable properties (source-dependent)
- Moisture sensitive (varies)

HOW PROPERTIES INFLUENCE MATERIAL SELECTION

Strength & Stiffness
Resist loads and limit deformation.

Thermal Properties
Withstand temperature changes and transfer heat as required.

Durability
Resist wear, corrosion, fatigue, and environmental degradation.

Weight
Affects efficiency, performance, and ease of use.

Cost & Availability
Impact feasibility, scalability, and sustainability.

KEY TAKEAWAY: Understanding the core properties of material types enables informed decisions to achieve safety, performance, reliability, and sustainability in engineering design.

Chapter 4: Types of Packaging

Packaging exists in layers, each serving a distinct role in guiding a product from manufacturer to consumer. By examining primary, secondary, and tertiary packaging, we gain insight into how designers balance protection, presentation, and efficiency at every stage of the supply chain. In this chapter, we will explore each packaging level in turn, illustrating the design considerations, material choices, and logistical functions that define them. Understanding these types of packaging enables packaging professionals to craft coherent systems that protect goods, optimize handling, and reinforce brand identity.

Primary Packaging

The primary package is the enclosure that makes direct contact with the product itself. It must satisfy the strictest requirements for safety, compatibility, and functionality, since any failure here imperils the product and the consumer. Consider a bottled beverage: the glass or PET container protects the liquid from contamination and light exposure, while the closure—crown cap or screw-top—ensures a reliable seal against leakage and oxidation. In food applications, primary packaging must comply with food-contact regulations, necessitating inert materials and hygienic designs free of crevices where bacteria could harbor.

Beyond protection and safety, primary packaging is the consumer's first tactile and visual experience. Its form factor influences shelf appeal, ergonomics, and portion control. A yogurt cup, for example, must be easy to hold, simple to open, and fit freezer trays if destined for frozen-dessert aisles. Flexible pouches, popular for snacks and sauces, merge convenience and lightweight transport: spouted pouches allow controlled pouring, while resealable zippers extend freshness after opening. When designing primary packaging, one asks: How will the consumer interact with this package? What opening force is appropriate for target demographics? How can the structure minimize material yet maintain integrity under mechanical stress?

Secondary Packaging

Secondary packaging assembles one or more primary units into a single grouping for display, merchandising, or intermediate transport. Although it does not contact the product directly, its design profoundly influences brand communication and handling efficiency. A folding carton that cradles four primary beverage bottles allows brands to extend graphics across a larger canvas, telling stories through color, typography, and imagery. Retailers rely on these cartons for shelf-ready packaging: tear-away perforations transform shipping cartons into instant displays, reducing labor at the point of sale.

In e-commerce, secondary packaging often doubles as shipping container: mailer boxes printed with brand elements create memorable unboxing experiences while eliminating separate shipping cartons. Here, the secondary package must resist crushing, puncture, and edge compression during shipment. Designers employ material testing—edge crush tests (ECT) and box compression tests (BCT)—to determine the corrugated board grade and flute profile needed to support expected loads. By integrating structural strength with graphic finesse, secondary packaging serves as both protective vessel and marketing medium.

Secondary packages also facilitate inventory management. Barcodes and RFID tags affixed to cartons enable automated scanning in warehouses, while clear window panels can reveal product inside, helping store associates quickly identify stock. When planning secondary packaging, professionals address questions of dimension standardization—fitting multiples of primary units without waste—stack strength for palletization, and the ease of opening for both retailers and consumers.

Tertiary Packaging

Tertiary packaging, or transit packaging, governs the bulk movement of products through distribution networks. Its objective is purely functional: to protect grouped secondary cartons during long-distance transport, expedite handling, and maximize container utilization. Pallets, stretch film, slip sheets, and unit-load devices comprise the core components of tertiary packaging systems.

A standard wooden pallet may carry up to 1,000 kilograms of product, but without proper stabilization—stringent banding, corner boards, and shrink wrap—the load risks collapsing under vibration or shock. Engineers model pallet patterns to optimize weight distribution and airflow, ensuring that refrigerated or frozen shipments maintain temperature uniformity. In automated warehouses, tertiary packaging must conform to conveyor widths, pallet jack dimensions, and forklift specifications; a mismatched pallet height can halt a production line.

Shrink-wrapped pallets combine flexibility and cost-efficiency: a few rolls of film can secure dozens of pallets per shift, while film tension and overlap determine resistance to shifting loads. For highly sensitive goods—electronics or glassware—engineers may incorporate void-fill materials, anti-static film, or vibration-damping pads within tertiary packaging to mitigate transport stresses. The choice of pallet material itself (wood, plastic, or corrugated plastic) balances durability, cleanliness, and recyclability.

Designing tertiary packaging extends beyond physical protection: it encompasses labeling for hazard communication, handling instructions, and compliance with international shipping regulations. Dangerous goods require UN-approved drums or boxes marked with specific hazard labels and packaging group indicators. Bulk shipments of pharmaceuticals or temperature-sensitive vaccines call for insulated containers and data loggers that track thermal history throughout transit. By rigorously defining the functional requirements for tertiary packaging, professionals ensure that products arrive intact, compliant, and ready for downstream distribution.

Integrating Packaging Levels

While each packaging type fulfills distinct roles, the true art lies in harmonizing them into an efficient, coherent system. A well-designed primary package that requires extra cushioning may force secondary packs to add protective inserts, increasing material usage and cost. Conversely, robust secondary and tertiary structures can allow lighter primary packages, reducing raw-material consumption and environmental footprint.

Consider a premium cosmetic set: a delicate glass bottle (primary) sits within a molded pulp tray (secondary) inside a branded rigid box. That box, in turn, is stacked on pallets with interlayers that support uniform load distribution (tertiary). At each level, designers ask whether functionality and aesthetics align with the brand's sustainability commitments. Could the molded pulp tray be sourced from post-consumer paper? Might the rigid box double as a gift box in the consumer's home? How can pallet configurations minimize empty space and optimize shipping container efficiency?

By mapping the entire packaging chain—from filling line to recycling bin—professionals identify synergies and trade-offs. Lifecycle assessments quantify environmental impact at each stage, revealing opportunities for lightweighting, material substitution, or reuse schemes. Collaborative workshops with suppliers, logistics partners, and recycling organizations ensure that packaging innovations succeed not just in the lab but across complex, real-world supply chains.

Conclusion

Distinguishing primary, secondary, and tertiary packaging clarifies the roles each plays in protecting products, communicating brand value, and enabling efficient distribution. As a packaging professional, your challenge is to orchestrate these layers into a unified system that meets functional requirements, aligns with economic targets, and advances sustainability goals. By mastering the design considerations at each level and fostering cross-functional collaboration, you will create packaging solutions that delight consumers, streamline operations, and stand up to the rigors of global commerce. In the next chapter, we will turn our attention to the principles of structural and graphic design, exploring the technical and aesthetic choices that bring packaging to life.

CHAPTER 4

Types of Packaging

Differentiate primary, secondary, and tertiary packaging levels.

Packaging is used at different levels to protect products, facilitate handling and transport, and communicate information throughout the supply chain.

PRIMARY PACKAGING
(Consumer Packaging)

The first level of packaging that directly contacts and contains the product.

PURPOSE

- Ensures product protection, safety, and hygiene
- Provides product information and branding
- Designed for individual sale or use

EXAMPLES: Bottles, cans, pouches, blister packs, jars, tubes

SECONDARY PACKAGING
(Grouping Packaging)

The second level of packaging that groups primary packages together.

PURPOSE

- Provides additional protection
- Facilitates handling, stacking, and storage
- Carries information for groups of products

EXAMPLES: Cartons, boxes, shrink wrap, trays, multipacks

TERTIARY PACKAGING
(Transport Packaging)

The third level of packaging used to handle, protect, and transport large quantities of products.

PURPOSE

- Protects products during transport and handling
- Enables efficient handling and distribution
- Supports bulk shipment and global logistics

EXAMPLES: Pallets, stretch wrap, crates, pallet boxes, shipping containers

AT A GLANCE

PRIMARY PACKAGING (Product Contact) → SECONDARY PACKAGING (Product Grouping) → TERTIARY PACKAGING (Bulk Transportation)

Each packaging level plays a critical role in delivering products safely, efficiently, and cost-effectively to the end user.

KEY TAKEAWAY: Understanding the three levels of packaging helps in selecting the right materials and designs to ensure protection, efficiency, and cost-effectiveness throughout the supply chain.

Chapter 5: Packaging Design Principles

Packaging is the bridge between product and consumer, marrying form and function to protect, communicate and persuade. In this chapter we explore the fundamental design principles that guide every successful package, from its internal structure to its outward appearance and the ways it serves human needs while meeting commercial constraints. By understanding these interlocking concepts, you will gain a holistic framework for creating packaging that performs reliably, resonates with brand values and delights users.

Structural Design Basics

At the heart of every package lies its structure, the physical skeleton that must withstand forces during transport, storage and handling. Finite element analysis (FEA) offers a computational approach to evaluate stress distribution, deformation and potential failure points under varied load conditions. For a beginner, it suffices to understand that FEA models approximate how materials bend, buckle or break when compressed, stacked or dropped. By simulating scenarios—such as pallet stacking or corner impacts—a designer can optimize wall thickness, corrugation profiles or reinforcement ribs to achieve the necessary strength without over-engineering.

Material selection plays a pivotal role in structural integrity. Paperboard, corrugated fiberboard, plastics, metals and composites each present unique stiffness, tensile strength and puncture resistance characteristics. When selecting a substrate, consider the environmental conditions the package will face—

humidity, temperature fluctuations and chemical exposure all influence material performance over time. A moisture-sensitive product transported through humid climates demands water-resistant coatings or liners, while a heavy glass container may benefit from high-strength corrugated board with reinforced flutes. Balancing mechanical performance against cost and sustainability targets is the designer's ongoing challenge: choose too lightly, and the package fails; choose too heavily, and you inflate expenses and environmental impact.

Beyond material and shape, closure systems and seals contribute to structural success. Snap-fit lids, tuck-tabs and heat seals must achieve both secure retention and user accessibility. The human hand applies variable force, so closures should incorporate features like tapered surfaces or interlocking geometries that guide assembly and disassembly without undue effort. Prototyping simple paper models or plastic mock-ups early in the process allows you to test closure performance before investing in tooling.

Graphic Design and Branding

While structure protects, graphic design communicates. Visual hierarchy directs the eye through typography, imagery and color, guiding consumers to essential information—brand name, product benefits, usage instructions. Establish a clear focal point, often the logo or headline claim, by contrasting size, weight or hue. Surrounding elements—descriptive text, ingredient listings, certifications—should recede subtly, maintaining legibility without competing for attention.

Consistency is the hallmark of strong branding. Packaging across a product line must share common visual elements—color palettes, typefaces, iconography and photographic style—so that consumers recognize your brand on crowded shelves. A cohesive identity fosters trust and loyalty. Yet consistency need not become monotony. Within established guidelines, you can introduce nuanced variations—accent colors or localized imagery—to tailor packaging for specific markets or seasonal promotions while remaining unmistakably part of the same family.

Color science underpins emotional resonance. Warm hues such as reds and oranges evoke energy and appetite, making them popular in food packaging, while cooler blues and greens suggest calm, health or eco-friendliness. Beyond psychology, practical considerations arise: vibrant inks may cost more or require multiple print passes, and certain substrates absorb pigments differently, altering the final appearance. Collaborate closely with your printer to understand ink coverage, dot gain and substrate gloss, ensuring your chosen palette reproduces accurately at scale.

Imagery and illustration further reinforce brand narratives. Photographic renders of the product in use create contextual relevance, while stylized illustrations can convey craftsmanship or heritage. Whether you use stock images or bespoke artwork, ensure resolution and file formats meet printing standards to avoid pixelation or color shifts. In an academic setting, consider creating mood boards to align stakeholders on the desired visual tone before moving into detailed dieline layouts.

User-Centered Design

Human factors must guide all design decisions. Ergonomics focuses on how people interact with packaging—lifting, opening, pouring or closing. Analyze the range of hand sizes, grip strengths and dexterity levels among your intended users. A child-proof cap that functions for adults may frustrate elderly consumers. Conversely, tear notches, perforations and pull-tabs should strike a balance between tamper evidence and user accessibility. Clear visual cues—icons or scored lines—help users instinctively know where to grip or tear.

Accessibility goes beyond physical interaction. Consider color-blindness when choosing color combinations for warnings or instructions, and ensure contrast ratios meet readability standards. For visually impaired users, embossing or tactile indicators can signal orientation—"lift here" or "this side up"—without relying solely on printed text. Packaging that accommodates diverse users not only expands your market but also demonstrates social responsibility.

User testing validates design assumptions. Even a simple paper prototype can reveal unforeseen friction points—users might struggle to open a flap or misinterpret closure mechanics. Organize small sessions where participants from your target demographic perform common tasks with the prototype while you observe and note difficulties. Iterative feedback loops, though time-consuming, prevent costly mistakes later in production and sharpen the user experience.

Cost and Efficiency Considerations

Every design innovation must live within commercial realities. Economies of scale drive per-unit pricing: higher tooling investment and larger print runs lower the cost per piece, but only if sales volume justifies them. Early in the concept phase, estimate expected production quantities and communicate with manufacturers to understand minimum order quantities, setup fees and material surcharges. A luxury brand with limited runs may accept higher per-unit costs, whereas a mass-market consumer good demands the leanest bill of materials.

Material minimization strategies reduce waste and expense. Right-sizing packages so they closely hug the product not only saves substrate but also lowers shipping volume and associated logistics costs. In the era of e-commerce, anti-slack design—filling empty space with molded inserts or inflatable cushions—protects products while keeping overall package dimensions lean. Revisit margin allowances regularly; as ingredient or component tolerances evolve, your snug-fit package may suddenly become too tight or too loose.

Beyond raw materials, consider the efficiency of printing and converting processes. Complex dielines with excessive scores or cutting paths increase machine cycle time, driving up labor and energy costs. Simplify panel layouts, combine printing steps where possible, and choose fold patterns that run smoothly on standard equipment. Align your design with the capabilities of your supplier's presses, avoiding exotic requirements that travel through multiple specialized vendors and inflate lead times.

Sustainable design often parallels cost savings. Replacing mixed-material laminates with mono-materials streamlines recycling, and uncoated boards eliminate the need for cumbersome separation at end-of-life. In many regions, regulatory pressures and brand commitments are accelerating the shift toward recyclable, compostable or reusable packaging. Stay informed about local recycling standards and consumer expectations; a sustainable choice can enhance brand reputation and preempt future compliance costs.

Conclusion

The principles outlined here—structural integrity, graphic clarity, user empathy and cost efficiency—form a cohesive framework for packaging design. Each element influences the others: a beautifully branded package fails its purpose if it collapses under load, just as a robust container with uninspired graphics may go unnoticed by consumers. By integrating mechanical analysis, visual strategy and human-centered research early in the process, you create packaging that not only protects and persuades but also builds trust and satisfaction. As you move forward, remember that simplicity often wins: convey your brand story and ensure reliable performance without unnecessary complexity. With these guidelines, you are equipped to tackle real-world packaging challenges and to innovate responsibly within the dynamic landscape of materials, technologies and consumer needs.

CHAPTER

5

Packaging Design Principles

Teach structural and graphic design fundamentals, build skills.

Effective packaging design combines structural integrity and visual communication to protect the product, attract the consumer, and support the brand.

STRUCTURAL DESIGN FUNDAMENTALS

Focus on the form, materials, and construction that protect the product, and function efficiently in the supply chain.

PROTECTION
Safeguards the product from damage, contamination, and environmental factors.

FUNCTIONALITY
Easy to open, use, reseal, store, and dispose of.

MATERIAL SELECTION
Choose the right material for strength, barrier, sustainability, and cost.

STRUCTURAL DESIGN
Use appropriate package forms, dimensions, and constructions.

MANUFACTURABILITY
Design for efficient production and consistent quality.

COMMON STRUCTURAL ELEMENTS

GRAPHIC DESIGN FUNDAMENTALS

Focus on visual communication that informs, persuades, and builds brand connection.

VISUAL HIERARCHY
Guide the consumer's eye to the most important information first.

TYPOGRAPHY
Use clear, readable fonts that reflect the brand personality.

COLOR
Use color to attract attention, convey meaning, and differentiate.

IMAGERY & GRAPHICS
Support the message, evoke emotion, and build brand recognition.

INFORMATION DESIGN
Communicate essential details clearly and truthfully.

KEY GRAPHIC ELEMENTS ON PACKAGING

DESIGN PROCESS

1. **RESEARCH** — Understand the product, user, market, and regulatory needs.
2. **IDEATE** — Generate concepts that solve problems creatively.
3. **DESIGN** — Develop structure and graphics using tools and prototypes.
4. **TEST** — Evaluate protection, usability, shelf impact, and sustainability.
5. **REFINE & PRODUCE** — Improve the design and prepare for scale in manufacturing.

GOOD DESIGN DELIVERS VALUE

- Protects the product
- Enhances user experience
- Builds brand preference
- Supports sustainability
- Drives business success

KEY TAKEAWAY: Strong packaging design integrates smart structure and compelling graphics to protect the product, communicate value, and create a lasting impression.

Chapter 6: Manufacturing Processes

From the germ of a packaging concept to the stack of finished cartons on a warehouse pallet lies a sequence of coordinated operations. In this chapter, we explore how raw materials are converted, shaped, and readied to hold products ranging from breakfast cereal to high-value electronics. Understanding these manufacturing processes gives designers and engineers a practical lens on feasibility, cost drivers, and quality considerations that shape every packaging decision.

Material Conversion Processes

At the heart of packaging manufacturing is the transformation of raw substrates into formats that can be further processed. Three fundamental conversion processes—paperboard folding, plastic extrusion, and molding—each address unique material characteristics and production volumes.

Paperboard folding is the cornerstone of many consumer-goods packages. A continuous roll of paperboard, often laminated or coated for moisture resistance, advances through a series of scoring and creasing stations. These stations precisely pre-crease lines where the material will fold, ensuring crisp corners and accurate dimensions. Subsequent folding platens or mechanical folders then bend the scored board into simple sleeves, cartons, or complex multi-piece designs. Parameters such as board thickness, fiber orientation, and moisture content must be tightly controlled; even slight variation can lead to poor fold definition, cracking, or misaligned flaps. For short runs or prototypes, hand-fed folder-gluers provide flexibility. In high-speed production, rotary folder-gluers operate at hundreds of meters per minute, feeding preprinted board through synchronized folding arms, gluing stations, and stacking conveyors.

Plastic extrusion transforms thermoplastic pellets into continuous film, sheet, or profile. In film extrusion, polymer granules—commonly polyethylene or polypropylene—enter a heated barrel, where rotating screws melt and homogenize the resin. The molten polymer is forced through a flat die or circular ring to form a sheet or blown film. Cooling rollers or air rings then solidify the film to the desired gauge. Film thickness, melt temperature, and line speed directly influence mechanical properties such as tensile strength and barrier performance. Extrusion coating applies a molten polymer layer onto a paperboard or foil substrate, bonding disparate materials into a multi-layer laminate. Extrusion processes demand careful temperature profiles to avoid gel formation or degradation, and precise die maintenance to ensure uniform gauge across the web.

Molding processes shape plastic, metal, or composite materials into discrete packaging components. Injection molding is ubiquitous for rigid containers, caps, and closures. Plastic pellets are heated and injected under high pressure into a closed mold cavity, where rapid cooling fixes the part geometry. Mold design complexity—from single cavity blocks to multi-cavity clusters—dictates cycle times and yield. For thin-walled packaging such as tub lids or custom trays, mold venting and cooling-channel layout are critical to prevent warpage and sink marks. Compression molding and transfer molding, more common for thermosets and composite parts, involve pre-measured charges of material that are pressed into heated molds, allowing precise control over fiber orientation and surface finish.

Each conversion process carries its own set of constraints—equipment capital cost, setup time, material waste, and achievable tolerances. A designer who understands these trade-offs can align packaging concepts with the most appropriate technology or iterate designs to reduce complexity and cost before tooling investments begin.

Forming and Shaping Techniques

Once a substrate exists in roll, sheet, or molded form, it must be cut, shaped, and assembled into its final geometry. Die-cutting, thermoforming, and injection molding (when used as a forming method rather than conversion) illustrate the spectrum of shaping technologies.

Die-cutting is the art of transforming flat sheet or web materials into intricate shapes. A die—composed of hardened steel blades arranged in the outline of the desired form—presses into the substrate to shear cutouts, create embosses, and apply crease lines in a single pass. Flatbed die-cutters offer extreme flexibility for short runs and prototypes, allowing quick die changes and manual feeding. In contrast, rotary die-cutters integrate cylindrical dies around a rotating mandrel, achieving speeds of several thousand sheets per hour. Die-cutting tolerances of ±0.1 mm are achievable with high-precision tooling and stable substrates, enabling detailed windows, locking tabs, and interlocking flaps. Stripping stations then remove waste matrix material, and gluing stations may apply adhesive for downstream assembly.

Thermoforming shapes heated plastic sheets into 3D trays, blisters, and clamshell packages. A sheet of PET, PVC, or PS is clamped over a mold, heated until pliable, and then drawn into or onto the cavity by vacuum, pressure, or mechanical assist. After cooling, trimmed edges yield the finished component. Thermoforming excels at creating deep, complex geometries with relatively low tooling cost compared to injection molding. However, draw depth and feature definition depend on sheet thickness, heating uniformity, and mold design. Plug assists can be used for heavy-draw areas to ensure material distribution and avoid thinning.

Injection molding can also serve as a forming technique when parts such as closures, spouts, or fastening features are directly molded onto preformed substrates. Known as overmolding or insert molding, this process places a prefabricated insert (for example a rubber gasket or paperboard label) into the open mold. Molten plastic is then injected around or through the insert, bonding the materials and forming a single, integrated component. Overmolding demands careful thermal management to protect the insert, and tooling design that ensures complete mold fill without damaging delicate features.

Selecting among these shaping techniques requires weighing factors such as tooling lead time, per-unit cost, design intricacy, and production speed. Designers often iterate prototypes with multiple methods—die-cut samples for concept review, thermoformed mock-ups for shelf visualization, and low-volume injection molds for final validation.

Filling and Sealing

After conversion and forming, empty packages must be filled with product and sealed to create a functional unit. Automated filling and sealing processes dominate high-volume lines, while semi-automatic or manual solutions serve niche and short-run requirements.

Automated filling lines handle liquids, powders, granules, and solids with specialized dosing equipment. Liquid products—beverages, condiments, or chemical solutions—may use piston fillers, volumetric pumps, or gravity fillers. Piston fillers deliver precise volumes by drawing fluid into a cylinder and then dispensing it into the container. Volumetric pumps maintain continuous throughput but require careful calibration when fluid viscosity varies. Powder and granule fillers rely on augers, vibratory feeders, or weigh scales to meter precise quantities into cartons, pouches, or rigid containers. Filling speed and accuracy depend on the product's flow characteristics and the container geometry. All filling operations must integrate sensors and feedback loops to detect misfills, jams, or foaming.

Sealing methods vary by package type. Paperboard cartons use hot-melt gluing or cold glue to secure flaps. Glue applicators deposit adhesive beads at specific locations, and folding rollers press the flaps until set. Flexible pouches and film-based packages often employ heat sealing, where two layers are pressed between heated bars to fuse them. Heat-seal temperature, dwell time, and pressure must be calibrated to the film's melting point and barrier requirements. Induction sealing creates a hermetic bond to metal foil liners within neck finishes, ensuring tamper evidence and shelf life. Snap-fit closures and crimped metals cans use mechanical sealing processes—press-on caps, crimping heads, or button seaming machines—to achieve leak-proof closures without adhesives.

Leak prevention is paramount. Visual inspection systems, pressure decay testers, and X-ray or ultrasonic inspection equipment detect improper seals, underfills, or foreign materials. Integrating these quality checkpoints in-line minimizes waste and recalls.

Manufacturing Process Flow: A Packaging Line in Action

Consider a mid-volume line producing six-pack carton packages for a bottled beverage. The journey begins with rolls of coated paperboard feeding a rotary die-cutter, which blanks and scores the flat carton blank. After stripping waste matrix, the blanks travel to a folder-gluer that applies adhesive and folds the boxes into nestable six-packs. Meanwhile, empty PET bottles arrive from a thermoforming and injection-molding department, then proceed to a rinsing station before entering a rotary filling carousel. Precisely metered piston fillers dispense the beverage, followed by a capping turret that applies screw caps and induction seals. Bottles exit this unit to a labeling operation, where adhesive labels are applied under controlled tension. Labeled bottles accumulate in a starwheel and index into the six-pack cartons, where a mechanical plunger inserts them into each compartment. The final sealing station folds and glues the carton's top flap; downstream, vision systems check each package for correct fill level, sealed flaps, and label placement. Completed cartons then travel on a conveyor to a palletizing robot, where they are stacked and wrapped for shipment.

This process flow illustrates the choreography of conversion, forming, filling, and sealing—each station optimized for speed, accuracy, and minimal changeover time. Effective line design considers footprint, material handling, labor requirements, and maintenance access. Early collaboration between designers and manufacturing engineers helps ensure that conceptual designs translate smoothly into high-throughput lines.

Conclusion

Understanding manufacturing processes transforms packaging from conceptual art into engineered reality. By mastering material conversion, shaping techniques, and filling and sealing operations, designers can anticipate production constraints, optimize cost, and preserve functionality and aesthetics. With this knowledge in hand, the next step is to explore quality assurance and performance testing—ensuring that every package not only looks right but delivers consistent protection and consumer satisfaction. In Chapter 7, we will delve into the metrics, standards, and inspection protocols that secure confidence throughout the supply chain.

CHAPTER

6

Manufacturing Processes

Explain key converting, forming, filling, and sealing methods.

Packaging is made through a series of processes that transform raw materials into finished packages, fill them with product, and seal them for protection.

CONVERTING

Transforms raw materials (films, paper, foil, metal, glass, etc.) into a form suitable for package production.

COMMON METHODS

Extrusion
Melting and shaping plastics into film, sheet, or profiles.

Slitting & Rewinding
Cutting large rolls into narrower rolls.

Printing
Applying text, graphics, and images.

Laminating
Bonding layers of different materials for performance or appearance.

Die Cutting
Cutting or creasing materials into specific shapes.

FORMING

Shapes the converted material into a package or container.

COMMON METHODS

Thermoforming
Heating plastic sheet and forming it over a mold.

Blow Molding
Inflating heated plastic to form hollow containers.

Injection Stretch Blow Molding (ISBM)
Stretching and blowing preforms to create strong, clear bottles.

Folding & Gluing
Folding flat materials (e.g., cartons) and applying adhesive to form boxes.

Metal Forming
Shaping metal sheets into cans, tubes, or other containers.

FILLING

Places the product into the package accurately, safely, and efficiently.

COMMON METHODS

Volumetric Filling
Delivers a specific volume (liquids, creams, pastes).

Gravimetric Filling
Fills by weight for high accuracy.

Auger Filling
Uses a screw conveyor to dispense powders or granules.

Piston Filling
Uses a piston to draw and dispense viscous products.

Flow Meter Filling
Measures flow rate for precise liquid filling.

SEALING

Closes the package and creates a barrier to protect the product and ensure integrity.

COMMON METHODS

Heat Sealing
Applies heat and pressure to seal films or laminates.

Induction Sealing
Uses electromagnetic energy to seal foil liners to containers.

Cap & Lid Sealing
Applies caps, lids, or closures using torque or pressure.

Crimping
Mechanically folds metal to secure closures (e.g., cans).

Adhesive Sealing
Uses glues or adhesives to close cartons or tubes.

TYPICAL PACKAGING PRODUCTION FLOW

1. **CONVERT** → Raw material is converted into usable webs or sheets.
2. **FORM** → Material is shaped into the package structure.
3. **FILL** → Product is accurately dispensed into the package.
4. **SEAL** → Package is sealed to ensure protection and quality.
5. **INSPECT & PACK** — Quality is checked, and products are prepared for distribution.

KEY POINTS

- Choosing the right processes ensures product protection, efficiency, and cost-effectiveness.
- Process selection depends on material type, product characteristics, package design, and production volume.
- Advances in automation and technology improve quality, speed, and sustainability.

KEY TAKEAWAY: Understanding converting, forming, filling, and sealing methods helps in selecting the right technology to create safe, functional, and appealing packages.

Better Packaging.
Better Future.

Chapter 7: Testing and Quality Assurance

Introduction

In the journey from concept to consumer, no stage is more critical to ensuring packaging performance than rigorous testing and quality assurance. For packaging students and professionals alike, understanding how physical stresses, environmental conditions, regulatory requirements, and systematic quality management converge is essential. This chapter examines four pillars of testing and assurance—physical and environmental testing, regulatory compliance, quality management systems, and an interactive QA audit checklist—so that you can design, evaluate, and validate packaging solutions that consistently protect products, satisfy stakeholders, and meet legal obligations.

Physical and Environmental Testing

Packaging must withstand the rigors of handling, transport, storage, and end-use. Physical and environmental tests simulate these real-world stresses in controlled laboratory settings to reveal vulnerabilities before packages ever leave the design table. Three of the most fundamental tests are drop testing, climate chamber exposure, and vibration testing.

Drop Testing

Drop testing replicates the impacts that occur when a package is inadvertently dropped or jostled during handling. Specimens are dropped from predetermined heights onto rigid surfaces, corners, edges or faces, following internationally recognized protocols. For example, a common standard for small shipments calls for a drop height of one meter onto each face. Observing how cushioning materials absorb shock, how seals hold, and whether contents shift or break provides immediate feedback on protective design. In practice, designers use drop test data to adjust material thickness, add internal supports, or change closure methods.

Climate Chamber Testing

Temperature and humidity extremes can degrade materials, warp shapes, and compromise barrier performance. Climate chambers subject packaging to cycles of heat, cold, and moisture—often spanning from -20 °C up to 60 °C with humidity control from 10 percent to 95 percent relative humidity. Over days or weeks, accelerated aging reveals how adhesives soften, plastics become brittle, and moisture-sensitive product integrity may be affected. For instance, food packaging designers monitor film permeability under high humidity to ensure microbial growth remains inhibited. Pharmaceutical packages undergo thermal cycling to confirm that blister seals resist delamination even when exposed to sudden temperature changes during global transit.

Vibration Testing

While drop testing addresses discrete shocks, vibration testing simulates continuous motion and resonance encountered during road, rail, sea, or air transport. Packages are secured to a vibrating table that oscillates at predefined frequencies and amplitudes. Engineers monitor for loosening of closures, abrasion between components, and eventual material fatigue. By mapping vibration profiles to specific shipping modes—such as a freight train's low-frequency rumble or an aircraft's high-frequency hum—designers tailor cushioning, alter mass distribution, or introduce shock-absorbing inserts. Vibration data thus informs both product protection strategies and decisions about pallet configuration.

Linking Test Data to Design Decisions

Interpreting results from physical and environmental tests is as important as running the tests themselves. A package that survives a single drop may still fail under repeated minor shocks; a climate-resistant film might react adversely when combined with an untested ink or adhesive. The academic rigor of standardized tests gives way to the creative adaptation of materials science—iterating prototypes, correlating laboratory failures with field reports, and optimizing designs for the full spectrum of handling environments.

Regulatory Compliance

Testing for durability addresses functional performance, but packaging professionals must also navigate a complex regulatory landscape. Whether your package holds food, pharmaceuticals, hazardous materials, or consumer goods, compliance with relevant standards and statutes is nonnegotiable. Three areas of particular concern are food contact regulations, child-resistant packaging requirements, and hazardous materials protocols.

Food Contact Materials

Packaging that touches or surrounds food must not transfer harmful substances or alter taste and aroma. In most jurisdictions, materials are classified and approved on the basis of migration testing, which measures the transfer of chemicals from substrate to food simulants under defined conditions. Plastic polymers, paperboard coatings, and adhesive residues each face separate migration limits. Designing compliant food packaging involves selecting approved ink formulations, barrier films, and coatings, then verifying performance through laboratory analysis. Failure to comply can lead not only to product recalls and brand damage but also serious health risks to consumers.

Child-Resistant Packaging

For products containing pharmaceuticals or hazardous substances, child-resistant closures reduce the risk of accidental ingestion. Legislation typically mandates dual-action mechanisms—such as push-and-turn caps—that are operable by adults but challenging for young children. Testing protocols involve panels of children and adults attempting to open the package within set timeframes. Designers must balance safety with usability, ensuring that elderly or mobility-impaired users can still access the product. Iterative rounds of prototype testing refine hinge stiffness, latch geometry, and instructional labeling to satisfy both performance criteria and user-experience standards.

Hazardous Materials Regulations

Transport of hazardous materials—flammable liquids, corrosive chemicals, compressed gases—falls under stringent international and national regulations. Packaging must be certified via tests such as hydrostatic pressure, drop tests at specified heights, stacking tests, and temperature exposure. Each package receives a UN performance code indicating material group, packaging type, and allowable contents. Noncompliant packaging risks shipment rejection, fines, or environmental harm. Professionals must remain current with updates to regulations like the UN Model Regulations, ADR (Europe), DOT (U.S.), and IATA (air cargo) regulations to ensure seamless global trade.

Integrating Compliance into Design Workflow

Regulatory requirements are not an afterthought; they shape material selection, closure mechanisms, labeling art, and documentation processes. Early collaboration with regulatory experts, material suppliers, and testing labs prevents costly redesigns. A holistic approach embeds compliance criteria within the conceptual phase, using checklists and decision trees so that functional, environmental, and legal demands are addressed in parallel.

Quality Management Systems

While testing and compliance identify individual concerns, a robust quality management system (QMS) ensures that all aspects of packaging design, production, and delivery adhere to consistent standards. Two widely adopted frameworks are ISO 9001 and Six Sigma—each offering a structured approach to continuous improvement and customer satisfaction.

ISO 9001 in Packaging

The ISO 9001 standard establishes requirements for a QMS focused on processes and documentation. For packaging, this encompasses incoming material inspections, production control procedures, calibration of test equipment, nonconformance handling, and customer feedback loops. By mapping processes and defining key performance indicators—such as defect rates, on-time delivery, and customer complaints—organizations can identify bottlenecks and opportunities for enhancement. Certified ISO 9001 systems signal to clients that the supplier adheres to internationally recognized quality practices.

Six Sigma Methods

Six Sigma complements ISO 9001 by providing data-driven tools for reducing variation and eliminating defects. Using the DMAIC (Define, Measure, Analyze, Improve, Control) roadmap, packaging teams tackle problems such as inconsistent glue application, dimensional tolerances that fall outside specification, or inefficiencies in warehousing. For example, a Six Sigma project might analyze the root causes of seal failures on a high-speed packaging line, implement corrective measures such as automated glue nozzles and real-time vision inspection, then establish control charts to maintain performance. By targeting a defect rate of 3.4 per million opportunities, Six Sigma drives both cost savings and customer satisfaction.

Cultivating a Culture of Quality

Beyond formal frameworks, quality assurance thrives when every team member—from design engineers to line operators—embraces ownership of outcomes. Regular training, open communication of test results, and recognition of improvement efforts foster an environment where potential failures are caught early. A strong QMS empowers organizations to respond quickly when anomalies occur, minimizing downtime and preserving brand reputation.

Interactive Checklist: QA Audit for a Fictional Beverage Package

To illustrate how the components of testing and quality assurance intersect, consider a fictional company preparing to launch a shelf-stable fruit juice in a 500 mL plastic bottle with tamper-evident cap. Use the following checklist as a guide for a comprehensive QA audit:

Physical Integrity

• Confirm drop test at 1.2 m without cracks or leaks.

• Validate vibration test matching refrigerated truck profile for 60 minutes.

1. Environmental Resistance

• Expose bottle to five 24-hour cycles in climate chamber (5 °C to 40 °C, 70 percent RH).

• Inspect label adhesion and clarity after humidity exposure.

2. Regulatory Compliance

• Verify polymer grade is approved for acidic food contact under relevant food safety standards.

• Ensure tamper-evident seal meets local child-resistant requirements if juice contains botanical extracts.

3. Production Quality Controls

• Review batch records for consistent fill volume ±2 mL tolerance.

• Check torque settings on closure line to guarantee ease of opening for adults and clear evidence of tampering.

4. Documentation and Traceability

• Confirm test reports are archived in the QMS database with traceable sample IDs.

• Validate calibration certificates for drop tester, torque tester, and climate chamber are current.

5. Continuous Improvement Actions

• Capture any deviations and assign corrective action owners.

• Schedule re-audit after implementing process improvements.

Conclusion

Testing and quality assurance are not isolated checkpoints but an integrated ecosystem that underpins every decision in packaging design and production. By mastering physical and environmental tests, navigating regulatory complexities, championing robust quality management systems, and employing practical audit tools, students and practitioners alike can elevate packaging from a mere container to a strategic asset. As you apply these principles, remember that rigorous evaluation breeds confidence—in your designs, in your processes, and ultimately in the end-product experience. With a systematic approach to testing and assurance, you safeguard not only the package and its contents, but also the trust of consumers and the reputation of your organization.

CHAPTER 7

Testing and Quality Assurance

Introduce testing protocols, compliance, and quality systems.

Testing and quality assurance ensure that packaging performs as intended, protects the product, complies with regulations, and meets customer expectations.

1. TESTING PROTOCOLS

Standard tests evaluate package performance under real-world conditions.

- **Compression Test** – Measures strength to withstand stacking loads.
- **Drop Test** – Evaluates package ability to withstand impact and shock.
- **Environmental Test** – Assesses performance under temperature, humidity, and other conditions.
- **Vibration Test** – Simulates transportation vibrations to check durability.
- **Seal Integrity Test** – Verifies seal strength and leak resistance.
- **Barrier & Permeability Test** – Measures protection against moisture, oxygen, light, and other factors.

2. COMPLIANCE

Packaging must meet applicable standards and regulations.

KEY AREAS

- **Food Contact Safety** – Ensures materials are safe for use with food and beverages.
- **Regulatory Requirements** – Comply with local and international laws (e.g., FDA, EU, REACH).
- **Sustainability Claims** – Validate claims related to recyclability, compostability, and recycled content.
- **Labeling & Marking** – Ensure accurate labeling, warnings, and traceability information.

EXAMPLES OF STANDARDS

- ISO 9001 (Quality Management)
- ASTM (Testing Standards)
- ISTA (Transportation Testing)
- FSC® (Sustainable Forestry)
- FDA 21 CFR (Food Contact)

3. QUALITY SYSTEMS

Structured systems drive consistent quality and continuous improvement.

QUALITY SYSTEM ELEMENTS

- **Quality Management System (QMS)** – Establish processes, responsibilities, and documentation to ensure quality.
- **Process Control** – Monitor and control critical steps to reduce variation and defects.
- **Supplier Management** – Evaluate and approve suppliers; ensure quality across the supply chain.
- **Corrective & Preventive Actions (CAPA)** – Identify issues, fix root causes, and prevent recurrence.
- **Internal Audits** – Regularly assess processes and compliance to drive improvement.
- **Training & Competency** – Ensure team members have the skills and knowledge to maintain quality.

4. QUALITY ASSURANCE BENEFITS

- **Protects the Product** – Ensures packaging performs throughout the supply chain and shelf life.
- **Ensures Compliance** – Meets legal requirements and industry standards.
- **Improves Efficiency** – Reduces waste, rework, and costs through process control.
- **Builds Customer Confidence** – Delivers reliable, high-quality packaging that strengthens brand reputation.
- **Supports Sustainability** – Validates environmental claims and promotes responsible packaging solutions.

THE QUALITY ASSURANCE CYCLE

PLAN – Define requirements and test plan. → **TEST** – Conduct tests and collect data. → **EVALUATE** – Analyze results and compare to criteria. → **IMPROVE** – Implement actions and monitor results. → **DELIVER QUALITY** – Ensure consistent, compliant, and reliable packaging.

KEY TAKEAWAY: Robust testing, compliance, and quality systems work together to ensure packaging safety, performance, and trust from production to the end user.

Better Packaging. Better Future.

Chapter 8: Sustainability and Regulations

In the unfolding narrative of packaging design, sustainability and regulation stand as twin pillars guiding innovation and responsibility. As packaging professionals and students, we confront a landscape where material choices and design decisions carry ecological weight, and where legal frameworks shape what is permissible, desirable, and ultimately successful. This chapter equips you with a conceptual toolkit for assessing environmental impact, adopting sustainable strategies, and navigating the complex web of regulations that govern packaging worldwide. By understanding life-cycle assessment, key design interventions, circular economy principles, and the contours of major regulatory systems, you will be prepared to lead projects that are not only commercially viable but also ecologically sound and legally compliant.

Environmental Impact of Materials

Every packaging solution begins with a material choice, and each material presents a distinct environmental profile. Life-cycle assessment (LCA) offers a structured way to evaluate impacts from cradle to grave: from raw material extraction through production, distribution, use, and end-of-life. Though the detailed methodologies of LCA may suit specialized consultancies, the core concepts are accessible to any designer.

We break LCA into four temporal phases:

1. Upstream impacts – Energy and resource demands involved in sourcing and processing raw materials. For example, the extraction of crude oil for plastics entails fuel consumption, land disturbance, and associated emissions.
2. Manufacturing impacts – Emissions, waste, and energy use in converting raw feedstocks into packaging formats. Paperboard production, for instance, consumes water and chemicals that require careful treatment.
3. Use phase – Impacts during the distribution, retail, and consumption stages. Heavier packages raise transport emissions; fragile designs contribute to product damage and waste.
4. End-of-life impacts – How a package's disposal or recovery influences environmental burdens. Landfilling, incineration, mechanical recycling, and composting each carry distinct footprints.

Two core metrics emerge in assessing these phases:

• Carbon footprint, capturing greenhouse gas emissions across the life cycle, and

• Waste metric, quantifying the mass or volume of materials entering disposal streams.

While precise quantification may require specialized software, designers benefit from a qualitative grasp: materials with high upstream energy intensity (e.g., certain plastics or metals) tend to have larger footprints, while those that impede recycling increase waste burdens. A packaging professional's early skill is to map these tendencies—identifying hotspots such as energy-intensive processing or difficult-to-separate multi-materials—and to deploy design strategies that mitigate them.

Sustainable Packaging Strategies

Armed with insights from life-cycle assessment, designers can adopt targeted strategies to reduce environmental impact. Three approaches—lightweighting, recyclability enhancement, and refill/reuse models—combine conceptual clarity with practical relevance.

Lightweighting

Lightweighting reduces material use while maintaining performance. Thoughtful gauge reduction, tailored material distribution, and structural optimization can shave grams from each package without compromising protection. Consider a molded pulp tray redesigned to employ thinner walls reinforced by ribbing; the result is comparable strength at lower material volume. Successful lightweighting demands rigorous testing to ensure no trade-off in safety or shelf appeal.

Recyclability Enhancement

Designing for recyclability means selecting mono-materials or easily separable composites, avoiding problematic additives, and communicating recycling instructions clearly. Shifting from mixed laminates to co-extruded films with compatible polymers simplifies processing at end-of-life. Labels and adhesives should be water-soluble or mechanically removable, allowing containers to flow through recycling streams without contaminant residues. While industry practices vary by region, a universal principle holds: the simpler and cleaner the material stream, the higher the recovery rates.

Refill and Reuse Models

Refillable packaging extends the life of primary containers, shifting environmental burdens to durable assets rather than single-use units. Common in bulk dispensing of detergents and beverages, refill systems require robust, easy-to-clean vessels and user-friendly dispensing mechanisms. Reuse models, such as rental crates for produce or standardized beverage bottles, depend on logistics networks for collection, cleaning, and redistribution. While initial infrastructure investments can be significant, the per-use environmental footprint often falls dramatically when compared to one-way packaging.

Each of these strategies carries design challenges: ensuring lightweight structures can withstand real-world stresses, aligning material streams with local recycling capabilities, or architecting refill networks that balance consumer convenience with operational feasibility. By assessing context—market dynamics, waste infrastructure, consumer behavior—designers can tailor the combination of approaches to yield maximum sustainability gains.

Recycling and the Circular Economy

Recycling, once framed as the silver bullet for waste, now sits within the broader vision of a circular economy—one that seeks to minimize resource extraction and eliminate waste by keeping materials in active use. Recycling itself encompasses mechanical and chemical processes. Mechanical recycling involves shredding and remelting plastics or pulping paper products; chemical recycling breaks down polymers to their monomers for repolymerization. Both routes demand clean, homogeneous input streams to maintain quality.

Beyond recycling, circularity introduces principles of design for disassembly, product-as-service models, and industrial symbiosis. Packaging designed for easy component separation—snap-fit closures, minimal adhesive zones, standardized fastening—facilitates the recovery of high-value materials. In industrial symbiosis, one company's waste becomes another's feedstock: textile offcuts may supply insulation materials; spent glass from beverage production can form glass wool for construction. These exchanges hinge on transparent material passports and collaborative supply chains.

A closed-loop system represents the ideal: packaging that returns to the manufacturer for cleaning, refilling, or remanufacturing; beverage bottles that cycle dozens of times before retirement. While closed loops thrive in localized contexts—bottle-deposit schemes in urban centers, crate-return networks in distribution hubs—their principles can inform broader supply chain strategies: defining take-back responsibilities, investing in reverse logistics, and fostering partnerships among brands, waste managers, and regulators. Even when full circularity is not yet feasible, aiming to keep materials in use for as long as possible reduces virgin resource demand and cuts waste.

Regulations and Standards

Sustainability in packaging does not occur in a vacuum; it is anchored by an evolving tapestry of regulations and industry standards. Regulatory regimes vary in scope and stringency, but most share common objectives: reducing packaging waste, promoting recycling, and encouraging producer responsibility.

The EU Packaging and Packaging Waste Directive

In Europe, the Packaging and Packaging Waste Directive establishes targets for recovery and recycling of packaging materials. It obliges member states to achieve defined rates—for instance, recovering at least 65 percent of total packaging waste and recycling 55 percent of it. Producers contribute to funding national packaging waste schemes, placing financial responsibility on those who place packaging on the market. In practice, this has driven innovation in lightweighting, mono-material design, and investment in recycling infrastructure.

Global Perspectives

Beyond Europe, nations have adopted diverse measures:

- Extended Producer Responsibility (EPR) schemes in Canada, Japan, and parts of the United States require producers to finance end-of-life management.

- Plastic taxes and levies in India and the United Kingdom penalize excessive single-use plastic.

- Deposit-refund systems in Scandinavia incentivize consumer returns of beverage containers by attaching a small surcharge redeemed upon deposit.

Regulations often set material-specific requirements—banishing certain hazardous additives, mandating minimum recycled content in plastic packaging, or restricting non-renewable materials in compostable claims. International standards, such as ISO 18601 on packaging and the environment, offer voluntary frameworks for eco-design, life-cycle assessment, and material management. Although not legally binding, adherence to ISO standards signals commitment to best practices and can streamline market entry in regions that reference these guidelines.

Navigating these frameworks demands that packaging professionals cultivate regulatory literacy: interpreting legislation, aligning design specifications with legal thresholds, and collaborating with compliance specialists. Early engagement with legal requirements shapes material sourcing, testing protocols, and supply chain agreements, reducing the risk of costly redesigns or market withdrawals.

Conclusion

As packaging students and professionals advance toward mastery, sustainability and regulation emerge not as burdensome constraints but as catalysts for innovation. By integrating life-cycle thinking into material selection, embracing lightweight, recyclable, and refillable solutions, and designing for circular systems, you contribute to a more resilient resource economy. Simultaneously, by mastering the regulatory landscape—from the EU's Packaging Waste Directive to global EPR schemes—you ensure that your designs meet legal obligations and reflect industry leadership. In the next chapter, we will delve into cost modeling and budgeting for sustainable packaging projects, translating these conceptual insights into financial frameworks that support both environmental stewardship and business success.

CHAPTER 8 Sustainability and Regulations

Explore environmental impact, circular economy, and regulations.

Sustainable packaging protects people and the planet while meeting performance, safety, and regulatory requirements throughout the product life cycle.

1. ENVIRONMENTAL IMPACT

Packaging has environmental impacts across its entire life cycle.

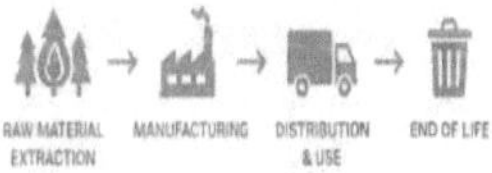

KEY IMPACT AREAS

Carbon Emissions
Contribute to climate change.

Resource Depletion
Consumes finite natural resources.

Waste Generation
Leads to landfill use and pollution.

Water & Ecosystem Impact
Affects water quality and biodiversity.

DESIGN FOR LOWER IMPACT

- Use fewer materials.
- Choose renewable or recycled content.
- Design for recyclability, reuse, or compostability.
- Optimize logistics to reduce emissions.

2. CIRCULAR ECONOMY

Keep materials and products in use for as long as possible to create value and reduce waste.

ENABLERS

- Collaboration across the value chain
- Infrastructure and collection systems
- Consumer education and participation
- Innovation and continuous improvement

3. REGULATIONS & COMPLIANCE

Packaging must comply with laws and standards that protect health, safety, and the environment.

Product Safety
Ensures packaging is safe for consumers and fit for its intended use.

Food Contact
Materials must meet food contact safety requirements.

Environmental Regulations
Address emissions, waste, recyclability, and restricted substances.

Extended Producer Responsibility (EPR)
Producers are responsible for the end-of-life management of packaging.

Labeling & Information
Requires accurate claims, instructions, and disposal information.

EXAMPLES

- **Global:** EU Single-Use Plastics Directive, ISO 14001, ISO 22000, GFSI, FSC®
- **Regional:** EU Packaging and Packaging Waste Regulation (PPWR), REACH
- **National:** FDA (USA), Health Canada, BIS (India)
- **Industry:** CPG guidelines, retailer standards

STAY COMPLIANT

- Understand applicable laws and standards.
- Monitor regulatory changes.
- Document and verify compliance.
- Train teams and partners.

SUSTAINABLE PACKAGING PRINCIPLES

Necessity
Only use packaging when it delivers clear benefits.

Reduce
Minimize material use and resource consumption.

Renewable
Use renewable materials and energy.

Recycle / Compost
Design for end-of-life recovery and return to the system.

Responsible
Act with transparency, ethics, and societal responsibility.

OUR GOAL
Deliver packaging solutions that are safe, compliant, ethical, and better for people and the planet.

KEY TAKEAWAY: Sustainable packaging balances performance, cost, and compliance while minimizing environmental impact through smart design, circular practices, and responsible choices.

Better Packaging.
Better Future.

Chapter 9: Careers in the Packaging Industry

Packaging has become an indispensable bridge between product and consumer, demanding ever more creative minds and technical expertise. For students embarking on this journey and professionals seeking advancement, understanding the breadth of opportunities and required competencies is crucial. In this chapter, we map the evolving industry landscape, explore the major sectors that drive demand for packaging talent, delineate the core roles and skills that define success, survey the educational and certification pathways available, and outline strategies to accelerate one's career. By the end, you will have clarity on where your interests may lie, how to acquire needed expertise, and the tactical steps to build a fulfilling career in this dynamic field.

Industry Landscape and Key Sectors

Food Packaging

The food sector remains the largest employer of packaging professionals, driven by global demand for convenience, extended shelf life, and sustainability. Innovations such as modified atmosphere packaging, active barrier films, and compostable trays have created roles for material scientists, process engineers, and sustainability analysts. Food-packaging teams collaborate with R&D chemists to balance cost, safety, and consumer appeal, ensuring freshness from farm to fork.

Pharmaceutical Packaging

Precision and compliance define pharmaceutical packaging. Regulatory requirements for child-resistant features, tamper-evident seals, and serialization protocols ensure patient safety and supply-chain security. Roles in this sector range from validation engineers who qualify clean-room processes to quality assurance specialists who audit batch records. The complexity of cold-chain logistics and biologics has expanded career avenues for thermal packaging experts and supply-chain analysts.

E-Commerce and Retail Packaging

The explosive growth of online commerce has placed new pressures on packaging to be protective, cost-efficient, and unboxing-worthy. Packaging designers work closely with fulfillment strategists to optimize parcel dimensions, reduce damage, and enhance the user experience. Data-driven roles—such as packaging performance analysts—emerge to interpret return rates and damage statistics. Cross-disciplinary collaboration between industrial designers and logistics managers is key to balancing brand storytelling with operational efficiency.

Industrial and Protective Packaging

Heavy machinery, automotive parts, and fragile electronics require bespoke protective solutions. Engineers specializing in corrugated design, foam inserts, and vibration analysis ensure that high-value goods survive harsh transit conditions. Career paths in this sector include structural design engineers who perform finite-element analysis, procurement specialists who negotiate material contracts, and environmental specialists who seek to minimize overpack and material waste.

Roles and Skills in Packaging

Research & Development

R&D professionals pioneer new materials, barrier solutions, and smart packaging technologies. They conduct laboratory tests on tensile strength, barrier performance, and biodegradability. Key skills include materials science fundamentals, experimental design, and familiarity with testing protocols such as ASTM or ISO standards. Strong communication is essential to translate lab breakthroughs into scalable production.

Packaging Design

Designers balance form and function, shaping the user experience while ensuring manufacturability. Proficiency in CAD software, structural design principles, and human-centered aesthetics is vital. Designers must also consider sustainability—selecting recyclable substrates or optimizing die-cut patterns to reduce scrap. Collaborative skills help them interface with marketing, production, and suppliers.

Production and Process Engineering

Process engineers set up and troubleshoot packaging lines, from filling machines to case erecting. They develop standard operating procedures (SOPs), perform line efficiency studies, and implement continuous-improvement methodologies like Six Sigma. Expertise in mechanical systems, automation controls (PLC programming), and statistical process control empowers engineers to enhance throughput and reduce downtime.

Quality Assurance and Regulatory Affairs

QA specialists enforce consistency and compliance, overseeing in-line inspections, sampling plans, and corrective-action investigations. Regulatory affairs professionals stay abreast of FDA, EU, and other global requirements, ensuring labeling accuracy, traceability, and child-resistance where mandated. Attention to detail, documentation mastery, and audit-management experience are hallmarks of success in these roles.

Supply Chain and Logistics

Supply-chain managers coordinate raw-material sourcing, inventory levels, and distribution strategies. They analyze lead times, negotiate vendor contracts, and build resilience against disruptions. Analytical skills, proficiency with ERP systems, and an ability to forecast demand help optimize costs while meeting service-level requirements. Sustainability is woven into modern supply chains, spawning roles in circular-economy planning and end-of-life management.

Education and Certification Paths

Packaging Degrees and University Programs

Universities worldwide now offer specialized degrees in packaging science, engineering, and technology. Curricula blend core courses in materials science, structural design, and supply-chain management with laboratory modules and industry internships. When selecting a program, consider accreditation by professional bodies and opportunities for hands-on capstone projects.

Professional Bodies and Certification

The Institute of Packaging Professionals (IoPP), the International Safe Transit Association (ISTA), and regional organizations provide certifications that validate expertise. The Certified Packaging Professional (CPP) credential from IoPP covers material science, design, distribution, and sustainability. Achieving certification typically requires a combination of education, work experience, and successful completion of a rigorous examination.

Continuing Education and Short Courses

Workshops on topics such as sustainable packaging design, advanced barrier coatings, and digital print technologies equip professionals to stay current. Many universities and private training firms offer online modules and weekend seminars. Engaging in these ensures that supply-chain changes, regulatory updates, and emerging materials are integrated into one's skill set.

Career Development Strategies

Building a Professional Network

Active involvement in industry conferences, webinars, and local packaging associations opens doors to mentors, collaborators, and job leads. Presenting case studies or technical papers elevates your profile and demonstrates thought leadership. Participating in online forums and LinkedIn groups focused on packaging allows knowledge sharing and visibility among peers and recruiters.

Crafting a Portfolio of Projects

A tangible portfolio that showcases design mock-ups, failure-mode analyses, or cost-optimization studies provides concrete evidence of your capabilities. Including before-and-after performance metrics—such as reduced carbon footprint or improved line efficiency—helps potential employers grasp the real-world impact of your work.

Continuous Learning and Innovation

The packaging industry moves rapidly, driven by environmental imperatives and consumer trends. Subscribing to trade journals, attending hackathons focused on sustainable materials, and experimenting with unconventional substrates (bioplastics, cellulose nanofibers) signal your commitment to innovation. Embrace cross-functional projects, whether in marketing or engineering, to broaden your perspective.

Navigating Career Transitions

Moving from one specialization to another—say, from production engineering into R&D—often requires bridging gaps through targeted certifications or short-term secondments. Express your interest in rotational programs within larger organizations, or seek mentorship arrangements that provide exposure to diverse functions. Clarity in your personal development plan, combined with measurable milestones, will accelerate these transitions.

Interactive Activity: Build Your Career Roadmap

Reflect on Your Interests and Strengths

List three aspects of packaging that excite you most—material innovation, structural design, regulatory affairs, or supply-chain optimization—and note any related projects you have completed.

Set Short-Term and Long-Term Goals

Define a one-year objective (for example, earning your CPP accreditation or leading a packaging line-efficiency study) and a five-year vision (such as becoming a packaging R&D manager or sustainability director).

Identify Learning Milestones

For each goal, determine the courses, certifications, or hands-on experiences required. Map out resources—online programs, professional associations, or internal training—to acquire those competencies.

Establish Accountability

Find a mentor in your organization or academic department to review your roadmap quarterly. Solicit feedback on progress and adjust your plan in response to emerging industry trends.

Conclusion

The packaging industry offers a spectrum of rewarding careers, from the laboratory bench to the global supply chain. By understanding the landscape, acquiring the right skills, and proactively shaping your professional journey, you can secure a role that aligns with your passions and talents. As you prepare to enter or advance within this field, let this chapter serve as your compass—guiding you toward opportunities where your contributions will protect products, delight consumers, and drive sustainable innovation. Next, we will explore the interplay of materials and sustainability, delving into how your career can leave a lasting positive impact on people and the planet.

CHAPTER 9

Careers in the Packaging Industry

Guide professional paths, required skills, and career planning.

The packaging industry offers diverse and rewarding career opportunities that combine creativity, technology, and sustainability to solve real-world challenges.

1. CAREER PATHS

Explore roles across the packaging value chain.

Packaging Design
Create innovative and functional package structures and graphics.

Materials Development
Develop and test new materials and sustainable solutions.

Production & Operations
Manage manufacturing processes, equipment, and efficiency.

Quality Assurance
Ensure product quality, testing, and compliance with standards.

Supply Chain & Logistics
Plan, coordinate, and optimize packaging supply and storage.

Sales & Marketing
Understand customer needs and communicate packaging value.

2. ESSENTIAL SKILLS

Build the skills that employers value.

Technical Knowledge
Materials, processes, equipment, and packaging standards.

Design & Creativity
Structural design, graphics, prototyping, and problem solving.

Analytical Thinking
Data analysis, testing, quality control, and process improvement.

Communication
Present ideas clearly and collaborate across teams and functions.

Project Management
Plan, execute, and deliver projects on time and within budget.

Sustainability Awareness
Understand environmental impact and circular economy principles.

Adaptability & Learning
Embrace new technologies and continuous improvement.

3. EDUCATION & EXPERIENCE

Prepare and grow your career.

EDUCATION PATHS

- High school: Build a strong foundation in math, science, and technology.
- Associate/Bachelor's degrees in Packaging, Engineering, Materials Science, Design, or related fields.
- Certifications and specialized courses add value.

GAIN EXPERIENCE

- Internships and co-ops
- Entry-level roles
- Cross-functional projects
- Professional development and continuous learning

GROW YOUR CAREER

- Develop expertise
- Take on leadership roles
- Expand your professional network
- Pursue advanced education or certifications

4. CAREER PLANNING TIPS

Take charge of your future.

SELF-ASSESS
Identify your interests, strengths, and values.

EXPLORE OPTIONS
Research roles, industries, and companies.

SET GOALS
Define short-term and long-term career goals.

BUILD YOUR BRAND
Create a strong resume, portfolio, and online presence.

NETWORK
Connect with professionals and join industry organizations.

STAY CURRENT
Follow industry trends, attend events, and keep learning.

PACKAGING: A WORLD OF OPPORTUNITIES

- Global industry with diverse sectors
- Purpose-driven work with real impact
- Collaborative and innovative environment
- Strong career growth and job stability
- Make a difference for people and the planet

WHERE CAN YOU WORK?

- Consumer Products
- Food & Beverage
- Pharmaceutical & Healthcare
- E-commerce & Retail
- Industrial & Transportation
- Sustainability & Recycling
- Government & Regulatory

KEY TAKEAWAY: With the right skills, mindset, and plan, you can build a fulfilling career and help shape the future of packaging.

Better Packaging. Better Future, Better Careers.

Chapter 10: Future Trends and Next Steps

As the packaging world accelerates into a new decade, professionals and students alike must understand the forces shaping tomorrow's solutions. This chapter explores four critical arenas—technological innovations, digital and smart packaging, global market dynamics, and actionable next steps—offering a roadmap to navigate emerging opportunities and challenges.

Technological Innovations

Advances at the material and process level are unleashing capabilities once confined to science fiction. Nanotechnology, for example, is enabling barrier coatings so thin they measure molecules by the billion, yet strong enough to repel oxygen and moisture at the atomic scale. Imagine a paperboard carton treated with a nanoparticle layer that extends fresh-food shelf life by days without compromising recyclability. Early adopters in pharmaceuticals are already trialing these coatings to stabilize sensitive compounds during transport.

Active packaging, too, leverages chemistry to interact dynamically with its contents. Oxygen scavengers integrated into film layers can neutralize residual air within vacuum-sealed pouches, while antimicrobial sachets embedded in perishables absorb pathogens before they spoil produce. Such systems shift packaging from passive containment to proactive preservation, offering brands a way to guarantee quality through complex logistics networks.

Three-dimensional printing rounds out this trio of breakthroughs. What once was a prototyping novelty now enables on-demand creation of bespoke package geometries optimized for product protection and material efficiency. Designers can iterate form, infill pattern and structural ribs in software, then print low-volume runs that rival injection-molded parts in performance. As printers grow faster and new polymers become printable, small-batch customization and rapid field-level adjustments will become commonplace.

Digital and Smart Packaging

In parallel with material advances, digitization is transforming packaging into an interactive interface. Radio-frequency identification (RFID) tags embedded beneath label layers allow real-time inventory tracking and anti-counterfeiting authentication. Major retailers use RFID to scan pallets and individual items at once, slashing labor costs and stock-outs. Meanwhile, QR codes printed directly on cartons grant consumers access to origin stories, nutritional data or loyalty programs with a simple smartphone scan.

The Internet of Things (IoT) pushes smart packaging further by integrating sensors that monitor temperature, humidity or shock. Cold-chain shipments of vaccines can now report their exposure history throughout transit, triggering alerts when thresholds are breached. Smart caps on beverage bottles can record opening time, empowering supply-chain partners to analyze consumption patterns and reduce spoilage. This connectivity not only enhances transparency but also fosters new business models—think conditional warranties activated only if transport conditions remain within spec.

Global Trends and Market Insights

Beyond individual technologies lies a macro landscape shaped by e-commerce growth, sustainability mandates and shifting consumer values. The rise of online retail has driven demand for protective yet lightweight mailers and modular box designs that ship flat and erect at fulfillment centers. Companies are investing in automation that erects, fills and seals packages in seconds, meeting the expectation of next-day delivery without ballooning costs.

Simultaneously, regulators and consumers are pressing brands to reduce plastic waste and carbon footprints. Extended producer responsibility laws in Europe and parts of Asia now hold manufacturers accountable for end-of-life materials, incentivizing recyclable, compostable or refillable architectures. Brands respond by exploring mono-material solutions—like paper-based pouches that replace multi-layer laminates—and by establishing take-back programs. In this environment, design decisions ripple across material markets, recycling infrastructures and consumer perceptions, underscoring the need for lifecycle thinking.

Next Steps and Resources

Armed with an awareness of these trends, readers can chart a course for professional growth and innovation. Begin by deepening technical expertise through specialized readings—titles on polymer chemistry, nanomaterial engineering and additive manufacturing will ground you in the science driving future packaging. Seek online courses offered by industry associations such as the Institute of Packaging Professionals, which convene experts and offer certifications in areas like sustainable packaging and supply-chain management.

Engagement with professional networks is equally vital. Attend conferences like PACK EXPO or interpack to witness pilot lines and live demonstrations of emerging technologies. Contribute to technical committees at ASTM or ISO, influencing the standards that will govern smart packaging protocols and recyclability guidelines worldwide.

Finally, cultivate a mindset of continuous experimentation. Set aside small-scale pilots in your organization or academic project to test active insert films or printed electronics. Document lessons learned, share findings in cross-functional forums, and iterate designs rapidly. In an industry propelled by both incremental refinements and disruptive leaps, agility and curiosity will be your greatest assets.

Looking ahead, the fusion of cutting-edge materials, interconnected systems and global imperatives will redefine what packaging can accomplish. By staying informed, building practical skills and contributing to the conversation, you position yourself not just to adapt but to lead the next wave of transformative packaging solutions.

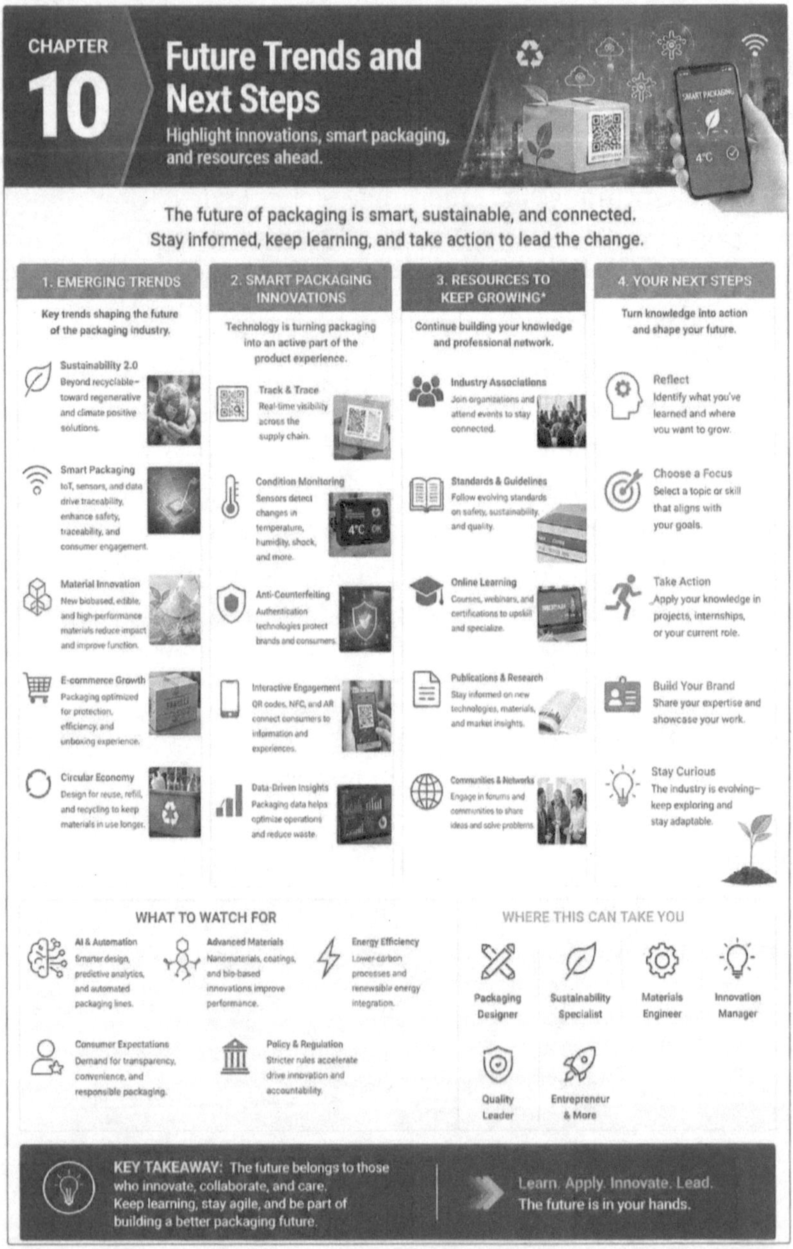
CHAPTER
10
Future Trends and Next Steps
Highlight innovations, smart packaging, and resources ahead.
SMART PACKAGING
4°C
The future of packaging is smart, sustainable, and connected.
Stay informed, keep learning, and take action to lead the change.
1. EMERGING TRENDS
Key trends shaping the future of the packaging industry.
Sustainability 2.0
Beyond recyclable–toward regenerative and climate positive solutions.
Smart Packaging
IoT, sensors, and data drive traceability, enhance safety, traceability, and consumer engagement.
Material Innovation
New biobased, edible, and high-performance materials reduce impact and improve function.
E-commerce Growth
Packaging optimized for protection, efficiency, and unboxing experience.
Circular Economy
Design for reuse, refill, and recycling to keep materials in use longer.
2. SMART PACKAGING INNOVATIONS
Technology is turning packaging into an active part of the product experience.
Track & Trace
Real-time visibility across the supply chain.
Condition Monitoring
Sensors detect changes in temperature, humidity, shock, and more.
4°C OK
Anti-Counterfeiting
Authentication technologies protect brands and consumers.
Interactive Engagement
QR codes, NFC, and AR connect consumers to information and experiences.
Data-Driven Insights
Packaging data helps optimize operations and reduce waste.
3. RESOURCES TO KEEP GROWING*
Continue building your knowledge and professional network.
Industry Associations
Join organizations and attend events to stay connected.
Standards & Guidelines
Follow evolving standards on safety, sustainability, and quality.
Online Learning
Courses, webinars, and certifications to upskill and specialize.
Publications & Research
Stay informed on new technologies, materials, and market insights.
Communities & Networks
Engage in forums and communities to share ideas and solve problems.
4. YOUR NEXT STEPS
Turn knowledge into action and shape your future.
Reflect
Identify what you've learned and where you want to grow.
Choose a Focus
Select a topic or skill that aligns with your goals.
Take Action
Apply your knowledge in projects, internships, or your current role.
Build Your Brand
Share your expertise and showcase your work.
Stay Curious
The industry is evolving–keep exploring and stay adaptable.
WHAT TO WATCH FOR
AI & Automation
Smarter design, predictive analytics, and automated packaging lines.
Advanced Materials
Nanomaterials, coatings, and bio-based innovations improve performance.
Energy Efficiency
Lower-carbon processes and renewable energy integration.
Consumer Expectations
Demand for transparency, convenience, and responsible packaging.
Policy & Regulation
Stricter rules accelerate drive innovation and accountability.
WHERE THIS CAN TAKE YOU
Packaging Designer
Sustainability Specialist
Materials Engineer
Innovation Manager
Quality Leader
Entrepreneur & More
KEY TAKEAWAY: The future belongs to those who innovate, collaborate, and care.
Keep learning, stay agile, and be part of building a better packaging future.
Learn. Apply. Innovate. Lead.
The future is in your hands.

Chapter 11: Limitations and Challenges in Packaging Design

Packaging design is often perceived as a creative and visually driven discipline; however, in reality, it is a complex engineering and strategic function shaped by numerous constraints. A successful packaging solution is rarely the most advanced or visually appealing option, but rather the one that achieves the best balance among performance, cost, sustainability, manufacturability, and consumer expectations. In practice, packaging professionals must continuously navigate a wide range of limitations while striving to deliver efficient and reliable solutions.

One of the primary challenges lies in material selection, as every packaging material comes with inherent trade-offs. Materials offering excellent barrier properties against oxygen, moisture, and light—such as multilayer laminates—are often expensive and difficult to recycle. On the other hand, lightweight materials help reduce transportation costs but may compromise strength and durability. Additionally, compatibility between the packaging material and the product is critical, particularly in industries like pharmaceuticals and chemicals, where interactions can lead to contamination or degradation.

Cost constraints further complicate packaging design decisions. Fluctuations in raw material prices, along with the high costs associated with tooling, molds, and printing processes, place constant pressure on designers to optimize expenses. In many cases, technically superior packaging concepts are not implemented because they are not commercially viable. This ongoing tension between innovation and affordability is a defining characteristic of the field.

Sustainability has emerged as a major concern in recent years, driven by environmental awareness and regulatory requirements. There is increasing demand for recyclable, biodegradable, and compostable packaging solutions, along with initiatives such as Extended Producer Responsibility (EPR). However, sustainable materials often present limitations in terms of barrier performance, durability, and cost. As a result, designers must carefully balance environmental responsibility with functional requirements, while avoiding the risk of superficial or misleading sustainability claims.

Another significant set of challenges arises from supply chain and logistics considerations. Packaging must be designed to withstand long-distance transportation, repeated handling, and exposure to varying environmental conditions such as temperature and humidity. It must also endure stacking loads during storage and transit. Achieving this level of durability while maintaining cost efficiency and optimizing space utilization requires careful engineering and testing.

Manufacturing constraints also play a crucial role in determining packaging feasibility. Designs must be compatible with existing filling lines, sealing technologies, and production speeds. Variations in material behavior and process tolerances can impact consistency and efficiency. A packaging solution that appears ideal during the design phase may encounter significant challenges when scaled up for mass production.

Regulatory and compliance requirements add another layer of complexity, particularly in sectors such as food, pharmaceuticals, and cosmetics. Packaging must meet strict standards related to labeling, safety, migration limits, and tamper evidence. These regulations often vary across regions, making it difficult to develop standardized global packaging solutions. Adapting to diverse regulatory environments requires both technical knowledge and strategic planning.

Consumer expectations have also evolved significantly, influencing packaging design in multiple ways. Today's consumers demand convenience features such as easy opening, reclosability, and portability, along with visually appealing and informative designs. Meeting these expectations while maintaining product protection and cost efficiency presents an ongoing challenge for designers.

In addition, packaging serves as a critical tool for branding and marketing. It must capture attention on crowded retail shelves, communicate brand identity, and convey essential product information within limited space. Designers are often required to deliver premium aesthetics under tight cost constraints, balancing creativity with practicality.

Testing and validation are essential to ensure packaging performance, but they are often limited by time and budget constraints. While tests such as drop testing, compression testing, climate simulation, and shelf-life studies provide valuable insights, they cannot fully replicate all real-world conditions. As a result, packaging development frequently relies on a combination of testing, experience, and informed assumptions.

Innovation in packaging introduces both opportunities and risks. New materials and formats can offer improved performance or sustainability benefits, but they also involve uncertainties related to scalability, cost, and acceptance within the supply chain. Resistance from internal stakeholders can further slow the adoption of innovative solutions, making it necessary to carefully balance progress with reliability.

A particularly critical challenge in modern packaging is the trade-off between recyclability and performance. Multilayer materials provide excellent protection but are difficult to recycle, whereas mono-material solutions are more environmentally friendly but often lack the required barrier properties. Resolving this conflict remains one of the most pressing issues in the industry today.

Globalization has also introduced the need to adapt packaging designs for different markets. Variations in climate, regulations, infrastructure, and consumer behavior require localized solutions, even for the same product. This creates a tension between the efficiency of standardization and the flexibility of customization.

Time-to-market pressures further intensify these challenges. Companies are expected to launch products quickly, leaving limited time for design optimization and thorough testing. This often results in compromises that may affect long-term performance or efficiency.

The rapid growth of e-commerce has added new dimensions to packaging design. Unlike traditional retail environments, e-commerce packaging must withstand higher levels of handling and impact while remaining lightweight and cost-effective. It must also provide a positive unboxing experience for consumers, increasing the complexity of design requirements.

Finally, limitations in waste management infrastructure, particularly in developing regions, affect the effectiveness of sustainable packaging solutions. Even when packaging is technically recyclable, inadequate collection, segregation, and processing systems may prevent it from being properly recovered, reducing its environmental benefits.

In conclusion, packaging design is a multidisciplinary field defined by continuous trade-offs and practical constraints. The goal is not to achieve perfection, but to develop optimized solutions that balance competing priorities. Successful packaging professionals are those who can navigate these challenges with a combination of technical expertise, strategic thinking, and real-world experience.

CHAPTER 11

LIMITATIONS AND CHALLENGES IN PACKAGING DESIGN

Packaging design is often perceived as a creative and visually driven discipline; however, in reality, it is a complex engineering and strategic function shaped by numerous constraints. A successful packaging solution is rarely the most advanced or visually appealing option, but rather the one that achieves the optimal balance among performance, cost, sustainability, manufacturability, and consumer expectations. In practice, packaging professionals must continuously navigate a wide range of limitations while striving to deliver efficient, reliable, and scalable solutions.

1. **MATERIAL SELECTION**
   Every packaging material presents inherent trade-offs. High-barrier materials like multilayer laminates offer excellent protection but are often expensive and hard to recycle. Lightweight materials reduce costs but may compromise strength and durability. Compatibility with the product is critical to prevent contamination or degradation.

2. **COST CONSTRAINTS**
   Fluctuations in raw material prices, along with high tooling, mold, and printing costs, put constant pressure on designers to optimize expenses. Technically superior concepts are often not adopted due to limited commercial viability.

3. **SUSTAINABILITY CHALLENGES**
   Growing demand for recyclable, biodegradable, and compostable packaging is driven by environmental awareness and regulations such as EPR. However, sustainable materials may have limitations in barrier performance, durability, or cost. Designers must balance environmental responsibility with functional requirements and avoid misleading claims.

4. **SUPPLY CHAIN AND LOGISTICS**
   Packaging must withstand transportation, handling, stacking loads, and exposure to temperature and humidity variations. Achieving durability while maintaining cost efficiency and optimizing cube utilization requires careful engineering and testing.

5. **MANUFACTURING CONSTRAINTS**
   Designs must be compatible with existing filling lines, sealing technologies, and production speeds. Variations in material behavior and process tolerances can impact consistency and efficiency. Scale-up from design to mass production often reveals unforeseen challenges.

6. **REGULATORY AND COMPLIANCE REQUIREMENTS**
   Packaging must meet strict standards for labeling, safety, migration limits, and tamper evidence. Regulations vary across regions, making global standardization difficult and requiring technical knowledge and strategic planning.

7. **CONSUMER EXPECTATIONS**
   Consumers expect convenience features (easy opening, reclosability, portability), clear information, visual appeal, and sustainability. Meeting these expectations while ensuring protection and cost efficiency is an ongoing challenge.

8. **BRANDING AND MARKETING DEMANDS**
   Packaging must stand out on crowded shelves, communicate brand identity, and convey key information in limited space. Designers must deliver premium aesthetics within tight cost constraints, balancing creativity with practicality.

9. **TESTING AND VALIDATION LIMITATIONS**
   Tests such as drop, compression, climate simulation, and shelf-life studies provide valuable data but cannot replicate all real-world conditions. Development relies on a combination of testing, experience, and informed assumptions.

10. **INNOVATION RISKS**
    New materials and formats can offer performance or sustainability benefits but come with uncertainties related to scalability, cost, and supply chain acceptance. Internal resistance can slow adoption, requiring a balance between innovation and reliability.

11. **RECYCLABILITY vs PERFORMANCE TRADE-OFF**
    Multilayer materials provide excellent protection but are difficult to recycle, while mono-material solutions are more eco-friendly but may lack required barrier properties. Resolving this conflict remains one of the most critical challenges.

12. **GLOBALIZATION AND MARKET DIVERSITY**
    Differences in climate, regulations, infrastructure, and consumer behavior require localized packaging solutions for the same product. This creates tension between the efficiency of standardization and the flexibility of customization.

13. **TIME-TO-MARKET PRESSURES**
    Companies are expected to launch products quickly, leaving limited time for optimization and thorough testing. This often results in compromises that may affect long-term performance or efficiency.

14. **E-COMMERCE CHALLENGES**
    E-commerce packaging must withstand higher handling and impact, remain lightweight and cost-effective, and deliver a positive unboxing experience—adding functional and experiential requirements to the design.

15. **WASTE MANAGEMENT INFRASTRUCTURE LIMITATIONS**
    Inadequate collection, segregation, and recycling infrastructure—especially in developing regions—may prevent technically recyclable packaging from being recovered, reducing its environmental benefits.

In conclusion, packaging design is a multidisciplinary field defined by continuous trade-offs and practical constraints. The objective is not to achieve perfection, but to develop optimized solutions that balance competing priorities. Successful packaging professionals are those who can navigate these complexities with a combination of technical expertise, strategic thinking, and real-world experience.

Chapter 12: Practical Insights and Real-World Case Studies

Coca-Cola's Lightweight PET Bottle

In 2015, Coca-Cola reduced each 500 mL PET bottle's weight by 17%. The result? Annual savings of 92,000 metric tons of plastic, lower transportation emissions, and no compromise in shelf stability. This illustrates how material innovation can yield cost, performance, and sustainability gains.

Tetra Pak's Aseptic Cartons

By combining paperboard with thin layers of aluminum and polyethylene, Tetra Pak created shelf-stable cartons that can store milk and juice for months without refrigeration. Benefits include logistic savings and reduced spoilage in regions without reliable cold chains. Limitations involve complex recycling processes requiring specialized facilities.

Amazon's Frustration-Free Packaging

To minimize waste and improve unboxing, Amazon partnered with brands to offer products in easy-open, 100% recyclable packaging. The program eliminated over 500 million shipping boxes in one year—cutting costs and customer frustration. However, initial design coordination proved challenging for small suppliers unfamiliar with new guidelines.

IKEA's Flat-Pack Strategy

IKEA's signature flat boxes optimize space in trucks and warehouses. Furniture is disassembled, nested, and packed with minimal filler. This reduces shipping costs by up to 50% and lowers carbon emissions per unit delivered. Challenges include added assembly time for consumers and the need for clear instruction manuals.

Practical Insights: Choosing the Right Packaging Method

When selecting a packaging solution, weigh these factors:

1. Product Characteristics: shelf life, fragility, regulatory requirements

2. Supply Chain Constraints: shipping distances, handling conditions, cold chain availability

3. Brand Positioning: premium vs. value, sustainability claims, aesthetic messaging

4. Cost Structure: material costs, manufacturing complexity, end-of-life disposal fees

5. Environmental Impact: recyclability, carbon footprint, reuse potential

A flowchart exercise: map product needs (e.g., oxygen sensitivity → barrier film; heavy weight → consider molded pulp trays) to arrive at a shortlist of materials and formats. Prototype and test under real-world conditions—drop tests, climate chambers, consumer feedback—to validate performance.

1 FLOWCHART EXERCISE

Map product needs (e.g., oxygen sensitivity → barrier film; heavy weight → consider molded pulp trays) to arrive at a shortlist of materials and formats.

PRODUCT NEED / CHALLENGE	FUNCTIONAL REQUIREMENT	POTENTIAL SOLUTION (MATERIAL / FORMAT)	SHORTLIST
OXYGEN SENSITIVITY →	HIGH BARRIER TO OXYGEN →	BARRIER FILM (e.g., EVOH / PLA co-extrusion) →	BARRIER FILM POUCH
HEAVY WEIGHT →	HIGH LOAD SUPPORT →	MOLDED PULP TRAYS →	MOLDED PULP TRAY
MOISTURE PRONE →	MOISTURE RESISTANCE →	COATED PAPER CUP / BOWL (water-based coating) →	COATED PAPER CUP

SHORTLIST OF MATERIALS & FORMATS

2 PROTOTYPE & TEST UNDER REAL-WORLD CONDITIONS

Prototype and test under real-world conditions—drop tests, climate chambers, consumer feedback—to validate performance.

DROP TESTS

Evaluate impact resistance and product protection.

CLIMATE CHAMBERS

Assess performance under temperature and humidity extremes.

CONSUMER FEEDBACK

Gather real user insights on usability, convenience and perception.

PERFORMANCE VALIDATION

Confirm that the packaging meets functional, sustainability and brand criteria.

**VALIDATED SOLUTION:
SUSTAINABLE PACKAGING THAT PERFORMS IN REAL LIFE**

Conclusion: Building Your Packaging Foundation

This chapter has introduced the fundamental concepts, materials, and real-world examples that define modern packaging. You've seen how protection, communication, convenience, and sustainability intersect to shape design decisions. As you progress through this book, each subsequent chapter will dive deeper into specialized topics—material science, graphic design, supply chain logistics, and sustainability frameworks—equipping you to craft innovative, effective packaging solutions. Keep these core ideas in mind: successful packaging is never an afterthought; it's an integrated strategy that supports product performance, brand promise, and environmental responsibility.

CHAPTER 12

PRACTICAL INSIGHTS AND REAL-WORLD CASE STUDIES

Applying the principles of sustainable packaging to real products, real constraints, and real results.

IN THIS CHAPTER

Practical takeaways from choosing sustainable packaging	Real-world case studies across industries	Testing, validation and lessons learned	Actionable insights you can apply today

Sustainable packaging is more than a material choice—it is a systems approach that considers the full life cycle, from sourcing and design to end-of-life. This chapter presents practical insights and real-world case studies that demonstrate how businesses are navigating challenges, embracing innovation, and achieving measurable results.

VALIDATED SOLUTION:
SUSTAINABLE PACKAGING THAT PERFORMS IN REAL LIFE

www.ingramcontent.com/pod-product-compliance
Lightning Source LLC
LaVergne TN
LVHW090329210926
848480LV00004B/1350